METHODS IN MOLECULAR BIOLOGY

Series Editor
John M. Walker
School of Life Sciences
University of Hertfordshire
Hatfield, Hertfordshire, AL10 9AB, UK

For further volumes:
http://www.springer.com/series/7651

The Nucleus

Second Edition

Edited by

Ronald Hancock

Laval University Cancer Research Centre-CRCHUQ Oncology, Québec, QC, Canada; Systems Biology Group, Biotechnology Centre, Silesian University of Technology, Gliwice, Poland

 Humana Press

Editor
Ronald Hancock
Laval University Cancer Research
    Centre-CRCHUQ Oncology
Québec, QC, Canada

Systems Biology Group
Biotechnology Centre
Silesian University of Technology
Gliwice, Poland

ISSN 1064-3745   ISSN 1940-6029 (electronic)
ISBN 978-1-4939-1679-5   ISBN 978-1-4939-1680-1 (eBook)
DOI 10.1007/978-1-4939-1680-1
Springer New York Heidelberg Dordrecht London

Library of Congress Control Number: 2014951833

Cover Image Caption: Water content in the nucleus of a HeLa cell; pixels containing 0 to 50 % water are yellow and those containing 51 to 100 % water as a linear scale from light to dark blue (see Chapter 12).

Printed on acid-free paper

Humana Press is a brand of Springer
Springer is part of Springer Science+Business Media (www.springer.com)

Preface

This volume presents detailed recently developed protocols ranging from isolation of nuclei to purification of chromatin regions containing single genes, with a particular focus on some less well-explored aspects of the nucleus.

The methods described include new strategies for isolation of nuclei, for purification of cell type-specific nuclei from a mixture, and for rapid isolation and fractionation of nucleoli. For gene delivery into and expression in nuclei, a novel gentle approach using gold nanowires is presented. The developing interest in analysis of specific regions of chromatin is illustrated by protocols for the isolation and structural and proteomic analysis of chromatin containing a single gene or containing newly synthesized DNA. A widely used method to purify chromatin regions is immunoprecipitation (ChIP), but during isolation chromatin structure may be modified by DNA damage response systems, and conditions which allow these artifacts to be avoided are described.

The concentration and localization of water and ions are crucial for macromolecular interactions in the nucleus, and a new approach to measure these parameters by correlative optical and cryo-electron microscopy is described. Similarly, redox conditions in the nucleus have been little explored, and a method to follow the redox dynamics of nuclear glutathione is an important step in this direction.

An important aspect of analyzing images of nuclear structures is the extraction of quantitative information, and this volume presents methods and software for high-throughput quantitative analysis of 3D fluorescence microscopy images, for quantification of the formation of amyloid fibrils in the nucleus, and for quantitative analysis of chromosome territory localization.

The friendly and timely collaboration of the contributors to this volume is greatly appreciated.

Québec, QC, Canada   *Ronald Hancock*

Contents

Contributors

FLORIAN ARNHOLD • *IUF – Leibniz Research Institute of Environmental Medicine at Heinrich- Heine-University Duesseldorf, Duesseldorf, Germany*

BAOYAN BAI • *Department of Radiation Oncology and Molecular Radiation Sciences, The Johns Hopkins University School of Medicine, Baltimore, MD, USA; Sidney Kimmel Comprehensive Cancer Center, The Johns Hopkins University School of Medicine, Baltimore, MD, USA*

AGATA BANACH-LATAPY • *UMR3348 "Genotoxic Stress and Cancer," Centre National de la Recherche Scientifique, Institut Curie, Orsay, France*

VINCENT BANCHET • *Laboratoire de recherche en Nanosciences EA 4682, UFR Sciences Exactes et Naturelles, Université de Reims Champagne Ardenne, Reims, France*

SASCHA BENEKE • *Institute of Veterinary Pharmacology and Toxicology/Vetsuisse, University of Zurich, Zurich, Switzerland*

HÉLÈNE BOBICHON • *CNRS UMR 7369, UFR Médecine, Université de Reims Champagne Ardenne and CHU de Reims, Reims, France*

HINRICH BOEGER • *Department of Molecular, Cell and Developmental Biology, University of California, Santa Cruz, CA, USA*

THOMAS BOUDIER • *Sorbonne Universités, UPMC Université Paris 06, Paris, France*

CHRISTOPHER R. BROWN • *Department of Molecular, Cell and Developmental Biology, University of California, Santa Cruz, CA, USA*

STEPHANIE D. BYRUM • *Department of Biochemistry and Molecular Biology, University of Arkansas for Medical Sciences, Little Rock, AR, USA*

SANDEEP CHAKRABORTY • *Department of Biological Sciences, Tata Institute of Fundamental Research, Mumbai, Maharashtra, India; Plant Sciences Department, University of California, Davis, CA, USA*

JULIEN COCHENNEC • *CNRS UMR7196, INSERM U565, Muséum National d'Histoire Naturelle, Paris, France*

DAVID CORTEZ • *Department of Biochemistry, Vanderbilt University School of Medicine, Nashville, TN, USA*

MICHÈLE DARDALHON • *UMR3348 "Genotoxic Stress and Cancer," Centre National de la Recherche Scientifique, Institut Curie, Orsay, France*

ALEXANDRE DAVID • *CNRS UMR-5203; INSERM U661; UM1; UM2, Institut de Génomique Fonctionnelle, Montpellier, France*

HUZEFA DUNGRAWALA • *Department of Biochemistry, Vanderbilt University School of Medicine, Nashville, TN, USA*

CHRISTOPHE ESCUDÉ • *CNRS UMR7196, INSERM U565, Muséum National d'Histoire Naturelle, Paris, France*

JULIAN A. ESKIN • *Department of Biology and Rosenstiel Basic Medical Science Research Center, Brandeis University, Waltham, MA, USA*

JOACHIM GRIESENBECK • *Lehrstuhl fürBiochemie III, Biochemie-Zentrum Regensburg (BZR), Universität Regensburg, Regensburg, Germany*

YASMINA HADJ-SAHRAOUI • *Laval University Cancer Research Centre-CRCHUQ Oncology, Québec, QC, Canada*

STEPHAN HAMPERL • *Lehrstuhl für Biochemie III, Biochemie-Zentrum Regensburg (BZR), Universität Regensburg, Regensburg, Germany*

RONALD HANCOCK • *Laval University Cancer Research Centre-CRCHUQ Oncology, Québec, QC, Canada; Systems Biology Group, Biotechnology Centre, Silesian University of Technology, Gliwice, Poland*

MARKUS HENGSTSCHLÄGER • *Institute for Medical Genetics, Medical University of Vienna, Vienna, Austria*

STEVEN HENIKOFF • *Basic Sciences Division, Howard Hughes Medical Institute, Seattle, WA, USA; Fred Hutchinson Cancer Research Center, Seattle, WA, USA*

MENG-ER HUANG • *UMR3348 "Genotoxic Stress and Cancer," Centre National de la Recherche Scientifique, Institut Curie, Orsay, France*

MELISSA S. JURICA • *Department of Molecular, Cell and Developmental Biology, University of California, Santa Cruz, CA, USA*

MIJEONG KANG • *Department of Chemistry, KAIST, Daejeon, South Korea*

KARL KATHOLNIG • *Institute for Medical Genetics, Medical University of Vienna, Vienna, Austria*

BONGSOO KIM • *Department of Chemistry, KAIST, Daejeon, South Korea*

MASATAKA KINJO • *Laboratory of Molecular Cell Dynamics, Faculty of Advanced Life Science, Hokkaido University, Sapporo, Japan*

MUGDHA KULASHRESHTHA • *Department of Biological Sciences, Tata Institute of Fundamental Research, Mumbai, Maharashtra, India*

MARIKKI LAIHO • *Department of Radiation Oncology and Molecular Radiation Sciences, The Johns Hopkins University School of Medicine, Baltimore, MD, USA*

NATHALIE LALUN • *CNRS UMR 7369, UFR Médecine, Université de Reims Champagne Ardenne and CHU de Reims, Reims, Cedex, France*

YUN WAH LAM • *Department of Biology and Chemistry, City University of Hong Kong, Kowloon, Hong Kong*

ZHOU FANG LI • *Department of Biology, South University of Science and Technology of China, Shenzhen, Guangdong, P.R. China*

FRANÇOIS LOLL • *CNRS UMR7196, INSERM U565, Muséum National d'Histoire Naturelle, Paris, France*

ISHITA MEHTA • *Department of Biological Sciences, Tata Institute of Fundamental Research, Mumbai, Maharashtra, India; UM-DAE-Centre for Excellence in Basic Sciences, Biological Sciences, Mumbai, Maharashtra, India*

JEAN MICHEL • *Laboratoire de Recherche en Nanosciences EA 4682, UFR Sciences Exactes et Naturelles, Université de Reims Champagne Ardenne, Reims, France*

ANNA VON MIKECZ • *IUF – Leibniz Research Institute of Environmental Medicine at Heinrich-Heine-University Duesseldorf Duesseldorf, Germany*

FRÉDÉRIQUE NOLIN • *Laboratoire de Recherche en Nanosciences EA 4682, UFR Sciences Exactes et Naturelles, Université de Reims Champagne Ardenne, Reims, France*

JEAN OLLION • *CNRS UMR7196, INSERM U565, Muséum National d'Histoire Naturelle, Paris, France*

DOMINIQUE PLOTON • *CNRS UMR 7369, UFR Médecine, Université de Reims Champagne Ardenne and CHU de Reims, Reims, France*

MARKO POGLITSCH • *Attoquant Diagnostics GmbH, Vienna, Austria*

B. J. Rao • *Department of Biological Sciences, Tata Institute of Fundamental Research, Mumbai, Maharashtra, India*

Andrea Scharf • *IUF – Leibniz Research Institute of Environmental Medicine at Heinrich- Heine-University Duesseldorf, Duesseldorf, Germany*

Florian A. Steiner • *Basic Sciences Division, Howard Hughes Medical Institute, Seattle, WA, USA; Fred Hutchinson Cancer Research Center, Seattle, WA, USA*

Alan J. Tackett • *Department of Biochemistry and Molecular Biology, University of Arkansas for Medical Sciences, Little Rock, AR, USA*

Sean D. Taverna • *Department of Pharmacology and Molecular Sciences, Johns Hopkins University School of Medicine, Baltimore, MD, USA*

Pavel Tchelidze • *CNRS UMR 7369, UFR Médecine, Université de Reims Champagne Ardenne and CHU de Reims, Reims, France*

Christine Terryn • *Plate-forme IBISA, Université de Reims Champagne Ardenne, Reims, France*

Manisha Tiwari • *Laboratory for Nano-Bio Probes, Quantitative Biology Center, OLABB, Osaka University, Suita, Japan*

Thomas Weichhart • *Institute for Medical Genetics, Medical University of Vienna, Vienna, Austria*

Laurence Wortham • *Laboratoire de Recherche en Nanosciences EA 4682, UFR Sciences Exactes et Naturelles, Université de Reims Champagne Ardenne, Reims, France*

Jonathan W. Yewdell • *Laboratory of Viral Diseases, National Institute of Allergy and Infectious Diseases, Bethesda, MD, USA*

Part I

Isolation of Nuclei

Chapter 1

Cell Type-Specific Affinity Purification of Nuclei for Chromatin Profiling in Whole Animals

Florian A. Steiner and Steven Henikoff

Abstract

Analyzing cell differentiation during development in a complex organism requires the analysis of expression and chromatin profiles in individual cell types. Our laboratory has developed a simple and generally applicable strategy to purify specific cell types from whole organisms for simultaneous analysis of chromatin and expression. The method, termed INTACT for Isolation of Nuclei TAgged in specific Cell Types, depends on the expression of an affinity-tagged nuclear envelope protein in the cell type of interest. These nuclei can be affinity-purified from the total pool of nuclei and used as a source for RNA and chromatin. The method serves as a simple and scalable alternative to FACS sorting or laser capture microscopy to circumvent the need for expensive equipment and specialized skills. This chapter provides detailed protocols for the cell-type specific purification of nuclei from *Caenorhabditis elegans*.

Key words Nuclei, Cell type, *Caenorhabditis elegans*, Expression profiling, Chromatin profiling, INTACT

1 Introduction

Most multicellular organisms are comprised of different tissues and cell types. The differentiation of a specific cell type from undifferentiated progenitors requires the expression of specific genes at specific time points. This is achieved by a combination of chromatin-based mechanisms involving transcription factor binding, nucleosome remodeling, deposition of histone variants, and post-translational histone modifications [1, 2]. As a consequence, each cell type is characterized by a specific chromatin landscape that gives rise to a specific gene expression signature. To understand cell type-specific function and differentiation, it is important to understand what chromatin changes underlie these processes and what expression profiles arise from them. However, these questions are intractable when studying whole organisms, as differences in expression and chromatin profiles between cell types are

Ronald Hancock (ed.), *The Nucleus*, Methods in Molecular Biology, vol. 1228,
DOI 10.1007/978-1-4939-1680-1_1, © Springer Science+Business Media New York 2015

marginalized in mixed populations. In recent years, several approaches have been developed to obtain pure populations of specific cell types or to extract RNA or DNA from specific cell types. These methods include the use of cell lines derived from specific tissues [3, 4], modified RNAs [5, 6], FACS-based approaches either for dissociated cells or nuclei [7–12], Dam-methylase expression in specific tissues [13], laser capture microscopy [14–18], and affinity-purification of nuclei [19–21].

Affinity-purification of nuclei is fast, simple and can be carried out without specialized equipment. The method, termed *isolation of nuclei tagged in specific cell types* (INTACT), relies on the expression of an affinity tag on the nuclear envelope specifically in the cell type of interest (Fig. 1).

The method was initially developed in our lab for *Arabidopsis thaliana* and we have since adapted it to *Caenorhabditis elegans* and *Drosophila melanogaster* [19–21]. Similar strategies have also been developed by other labs to purify nuclei from *D. melanogaster* and *Xenopus* [22, 23]. The method allows for the simultaneous isolation of RNA and chromatin, thus allowing comparison of gene expression profiles directly to the underlying chromatin landscapes. We use a two-component system for nuclear tagging where a nuclear pore fusion protein serves as a substrate for

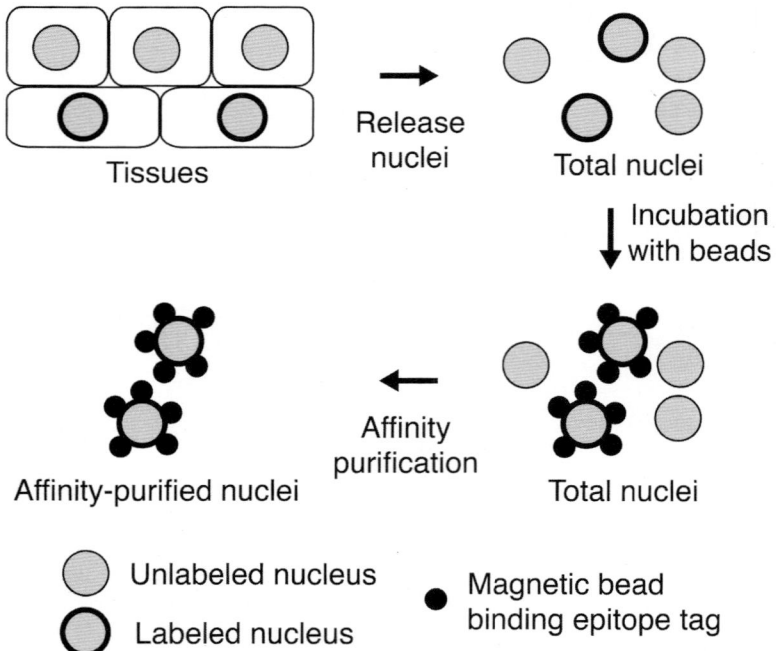

Fig. 1 Scheme of tissue-specific nuclei purification. Nuclei are epitope-labeled by the expression of a nuclear-tagging fusion protein specifically in the tissue of interest. Total nuclei are released and mixed with magnetic beads which recognize the epitope tag. The bead-bound nuclei are subsequently affinity-purified

biotinylation by *E. coli* biotin ligase (BirA), which is co-expressed in the same cells to mediate specific biotinylation. For *C. elegans*, we selected the outer nuclear pore protein NPP-9 and fused it to a tagging cassette that includes mCherry for visualization, biotin ligase recognition peptide (BLRP), a preferred substrate for BirA, and 3xFLAG for immunodetection. We call the NPP-9 fusion protein nuclear tagging fusion (NTF). We express a BirA::GFP fusion ubiquitously using the *his-72* promoter to enable biotinylation of the NTF in vivo [21].

Application of the method to purify muscle nuclei from adult *C. elegans*, using the *myo-3* promoter to express tagged NPP-9 in muscle, resulted in yields of 1–2 million nuclei with >90 % purity. Analysis of these nuclei revealed hundreds of genes that were specifically upregulated in muscle tissue, but also showed that the nucleosome occupancy was reduced over the promoters and within the bodies of these genes. The method also greatly increased the sensitivity of detecting changes in gene expression upon knock-down of the muscle-specific transcription factor HLH-1, underlining the importance of analyzing pure populations of nuclei when analyzing changes that only affect a subset of cells within an organism [21].

Here we provide a detailed protocol for the purification of tissue-specific nuclei from *C. elegans*. We divide the protocol into three stages: (1) *C. elegans* culture and fixation; (2) Isolation and affinity-purification of nuclei; and (3) Assessment of quality. The purified nuclei can be used as a source of RNA for gene expression profiling, of chromatin for micrococcal nuclease- or sonication-mediated fragmentation and chromatin immunoprecipitation, or of proteins for proteomic profiling.

2 Materials

2.1 Culture and Fixation of C. elegans Adults

1. Peptone-rich agar plates, 150 mm diameter: 12.5 g agar, 10 g peptone, 0.6 g NaCl, 1.5 g KH_2PO_4, 0.25 g K_2HPO_4, H_2O to 1 L.

2. *E. coli* strain NA22. Available from the Caenorhabditis Genetics Center (CGC), http://www.cbs.umn.edu/research/resources/cgc.

3. *C. elegans* strain expressing both NTF and BirA, e.g., strain JJ2300 expressing NTF in body wall muscle cells and BirA ubiquitously, available from the CGC.

4. *C. elegans* strain expressing only NTF (negative control), e.g., strain JJ2286 expressing NTF in body wall muscle cells, available from the CGC.

5. M9 solution: 22 mM KH_2PO_4, 42 mM Na_2HPO_4, 86 mM NaCl, 1 mM $MgSO_4$.

6. Phosphate-buffered saline (PBS): 8 mM Na_2HPO_4, 1.46 mM KH_2PO_4, 137 mM NaCl, 2.7 mM KCl, pH 7.4.

7. 50-mL conical tubes.

8. Sodium hypochlorite solution, 10–15 % available chlorine (Sigma).

9. 5 M NaOH.

10. N,N-dimethylformamide (Sigma).

2.2 Isolation and Affinity-Purification of Nuclei

1. Nuclei purification buffer (NPB): 10 mM Tris pH 7.5, 40 mM NaCl, 90 mM KCl, 2 mM ethylenediaminetetraacetic acid (EDTA), 0.5 mM ethylene glycol-bis(2-aminoethylethes)-$N_1N_1N;N^1$– tetraacetic acid (EGTA), 0.5 mM spermidine, 0.2 mM spermine, 0.2 mM DTT, 0.1 % Triton X-100.

2. NPB-0.5: NPB supplemented with Triton X-100 (0.5 % v/v).

3. Ceramic mortar and pestle.

4. Liquid nitrogen.

5. Glass dounce homogenizer, 7 mL.

6. Refrigerated tabletop centrifuge.

7. Bioruptor sonicator (Diagenode).

8. Refrigerated centrifuge for 50-mL conical tubes with swinging-bucket rotor (e.g., Sorvall Instruments RC5C).

9. Streptavidin M-280 Dynabeads (Invitrogen).

10. End-over-end rotator for Eppendorf tubes (e.g., Labquake Shaker, Thermo Scientific).

11. MiniMacs magnet (Miltenyi Biotec).

12. 10-mL serological pipettes.

13. NPB supplemented with 1 % (w/v) bovine serum albumin (BSA).

14. 1-mL micropipette tips.

15. Tygon tubing.

16. Hoffman tubing clamp.

2.3 Quality Control and Estimation of Yield

1. 4′,6-diamidino-2-phenylindole (DAPI).

2. Glass coverslips, 22×40 mm, No 1.5.

3. Microscope slides 3 well, 25×75 mm.

4. Upright fluorescence microscope (e.g., Zeiss Axioplan).

5. SDS-PAGE gel loading buffer (2×): 10 mM Tris–HCl, pH 6.8, 4 % (w/v) SDS, 20 % (v/v) glycerol, 2 % (v/v) β-mercaptoethanol, 0.04 % (w/v) bromophenol blue.

6. 100 °C heat block.

7. 6 and 18 % Tris–Glycine polyacrylamide gels.

8. Western blot transfer system.

9. Nitrocellulose membranes.

10. 2 % (w/v) bovine serum albumin (BSA) in PBS.

11. Anti-histone H3 C-terminus antibody (Abcam).

12. Anti-FLAG M2 antibody (Sigma).

13. Horseradish peroxidase (HRP)-conjugated streptavidin.

14. HRP-conjugated anti-rabbit and anti-mouse secondary antibodies.

15. Autoradiography film (Kodak).

3 Methods

3.1 Culture and Fixation of Adult C. elegans

1. Seed peptone-rich plates with 2 mL of an overnight culture of *E. coli* NA22. Incubate the plates overnight at 37 °C, then overnight at room temperature.

2. Grow the worm strain expressing both NTF and BirA (sample) and the strain expressing only NTF (negative control) on these plates until they are almost starved (*see* **Note 1**).

3. Wash the worms off the plates with M9 solution.

4. Wash the worms three times in M9 and resuspend them in 10 mL of M9 in a 50-mL conical tube.

5. Add 1 mL of sodium hypochlorite solution and 1 mL of 5 M NaOH.

6. Incubate with occasional shaking until the worm bodies have mostly disappeared, leaving behind embryos.

7. Pellet the embryos by centrifugation at $2,000 \times g$ for 1 min.

8. Wash the embryos three times with M9.

9. Plate the embryos on fresh NA22 plates (*see* **Note 2**).

10. Grow synchronized cultures to the young adult stage.

11. For one preparation, use worms from 2 to 4 plates or about 1–2 mL worm pellet, 1.5–2 million worms (*see* **Note 3**).

12. Wash the worms off the plates with M9 and collect in 50-mL conical tubes.

13. Pellet the worms to the bottom of the tubes by incubation on ice for 10–15 min.

14. Wash the worms 3–4 times with M9 and twice with PBS.

15. Centrifuge at $1,000 \times g$ for 2 min, remove residual PBS.

16. Add *N,N*-dimethylformamide cooled to −20 °C to a volume of 50 mL (*see* **Notes 4** and **5**).

17. Incubate at room temperature for 1 min (*see* **Note 6**).

18. Pellet the worms by centrifugation at $1,000 \times g$ for 1 min.

19. Remove the *N,N*-dimethylformamide and wash the worms three times with PBS cooled at 4 °C.

20. Freeze the worms dropwise in liquid nitrogen (*see* **Note 7**).

21. Grind the worms to a fine powder under liquid nitrogen.

3.2 Isolation and Affinity-Purification of Nuclei

1. All subsequent steps are carried out on ice or in the cold room. Buffers and equipment are precooled.

2. Add NPB to the worm powder to a total volume of 6 mL.

3. Transfer the suspension to a glass dounce homogenizer.

4. Break up the tissues with 30 strokes of the tight-fitting piston.

5. Distribute into six 1.5-mL centrifuge tubes.

6. Pellet debris by centrifugation at $100 \times g$ for 2 min in a refrigerated tabletop centrifuge.

7. Collect the supernatants containing nuclei and pool in a 50-mL conical tube.

8. Resuspend each pellet in 1 mL of NPB.

9. Sonicate with a Bioruptor sonicator at the lowest power output setting (130 W) twice for 30 s to release more nuclei (*see* **Notes 8** and **9**).

10. Pellet debris by centrifugation at $100 \times g$ for 2 min in a refrigerated tabletop centrifuge.

11. Collect the supernatants containing nuclei, and pool in the same 50-mL conical tube as above.

12. Repeat the sonication, low speed centrifugation and supernatant collection two more times.

13. Discard the pellets, which contain worm fragments and debris (*see* **Note 10**).

14. Bring the volume of nuclei collected in the 50-mL conical tube to 50 mL with NPB.

15. Pellet residual debris by centrifugation at $100 \times g$ for 5 min in a refrigerated centrifuge, then transfer the supernatant to a new 50-mL conical tube, discarding the pellet.

16. Add a 3-mL cushion of OptiPrep at the bottom of the tube (*see* **Note 11**).

17. Collect the nuclei as a layer on the cushion by centrifugation at $1,000 \times g$ for 10 min in a refrigerated centrifuge (*see* **Note 12**).

18. Transfer the nuclei into new 50-mL tube (*see* **Note 13**).

19. Bring the volume to 50 mL with NPB, add an OptiPrep cushion, and collect the nuclei on the cushion by centrifugation at $1,000 \times g$ for 10 min in a refrigerated centrifuge. Repeat for a total of three centrifugation steps (*see* **Note 14**).

20. Collect the nuclei in a 15-mL conical tube.

21. Retain 5 % of the sample for Western blot analysis and 1 % for microscopy (*see* Subheading 3.3, **steps 1** and **6**).

22. Bring the volume to 5 mL with NPB.

23. Add 30 μL of washed magnetic streptavidin-coated Dynabeads (*see* **Notes 15** and **16**).

24. Incubate for 45 min at 4 °C using an end-over-end rotator.

25. Bring the volume to 10 mL with NPB-0.5 (final concentration of Triton X-100 is 0.3 %).

26. Incubate 1-mL micropipette tips in NPB supplemented with 1 % (w/v) BSA (*see* **Note 17**).

27. Draw the nuclei into a 10-mL serological pipette.

28. Insert the pipette into a 1-mL micropipette tip that is wedged into a MiniMacs magnet. A schematic view of the assembled column is shown in Fig. 2.

29. Run the nuclei through the column at a flow rate of 1 mL/min. Control the flow rate with a Hoffman tubing clamp.

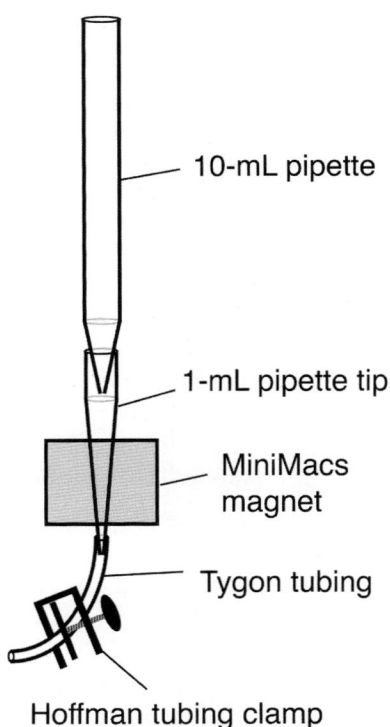

10-mL pipette

1-mL pipette tip

MiniMacs magnet

Tygon tubing

Hoffman tubing clamp

Fig. 2 Column setup for affinity purification. A 10-mL serological pipette is inserted into a 1-mL pipette tip which has been wedged into a MiniMacs magnet. A piece of Tygon tubing is attached to the pipette tip, and flow is controlled by a Hoffmann tubing clamp

30. Elute the beads into 10 mL of NPB-0.5.

31. Run the eluate through a second column (*see* Subheading 3.2, **steps 26–30** and **Note 18**).

32. Elute the beads into 500 μL of NPB.

33. Retain 5 % of the eluate for Western blot analysis and 1 % for microscopy (*see* Subheading 3.3, **steps 1** and **6**).

34. The remaining nuclei can be used for RNA, chromatin or protein isolation.

3.3 Quality Control and Estimation of Yield

1. To 1 % aliquots collected at **steps 21** and **33** of Subheading 3.2, add DAPI to a final concentration of 1 μg/mL.

2. Place a 1-μL aliquot of this sample onto a 3-well microscope slide.

3. Cover with a coverslip.

4. Count bead-bound vs. non-bead-bound nuclei under a compound microscope (*see* **Note 19**).

5. Compare to the number of nuclei in the negative control.

6. Mix 5 % aliquots collected at **steps 21** and **33** of Subheading 3.2 with equal amounts of SDS-PAGE sample buffer.

7. Incubate at 100 °C for 5–10 min.

8. Resolve the samples on two 6 % SDS-PAGE gels (one for detection of the FLAG epitope and one for detection of the biotinylated BLRP) and on one 18 % gel (for detection of histone H3).

9. Transfer the proteins to nitrocellulose membranes using a Western blot transfer system.

10. Block the membranes with 2 % (w/v) BSA in PBS (*see* **Note 20**).

11. Detect biotinylated BLRP with streptavidin-HRP, FLAG with anti-FLAG antibody followed by anti-mouse-HRP antibody, and histone H3 with anti-H3 antibody followed by anti-rabbit-HRP antibody (*see* **Note 21**).

12. Expose the membranes to autoradiography film.

4 Notes

1. We use a negative control for every purification to assess the quality of the pull-down. Nuclear lysis leads to clumping of the nuclei that can subsequently stick to the magnetic beads and lead to a large number of false positive nuclei. This problem is most easily recognized when a negative control is used. The problem of false positives caused by nuclear lysis and clumping is best addressed by gentler handling of the nuclei.

2. Treating adult worms with sodium hypochlorite and letting embryos hatch on fresh NA22 plates generates age-synchronized populations. This is important, as cells from the same tissue will have different expression and chromatin profiles depending on the stage of development.

3. The amount of worms needed to prepare nuclei of a given cell type will vary greatly depending on the required yield, the abundance of the cell type, and the efficiency of release of these nuclei from surrounding tissue. The first and third factors need to be determined empirically.

4. The purification can be done without fixation, as we have demonstrated for *D. melanogaster* mesoderm nuclei [21]. For native purifications, we omit detergents from the NPB to prevent lysis of the nuclei and use HB125 buffer (0.125 M sucrose, 15 mM Tris, pH 7.5, 15 mM NaCl, 40 mM KCl, 2 mM EDTA, 0.5 mM EGTA, 0.5 mM spermidine, 0.15 mM spermine, Roche Complete protease inhibitor cocktail) instead of NPB. However, when using the *C. elegans* NPP-9 tag we found that avoidance of any fixative leads to lower yields, possibly due to dissociation of NPP-9 from the nuclear pore.

5. 1 % formaldehyde can be used in the place of DMF for fixation. In this case, the 10 mM Tris is replaced by 50 mM HEPES in the NPB buffer because Tris quenches formaldehyde. We recommend the use of formaldehyde cross-linking if chromatin immunoprecipitation is the downstream application.

6. It is important to fix the worms lightly, as over-fixing will hinder the liberation of nuclei from the surrounding tissue.

7. Worms can be kept at −80 °C for a few days. However, we have experienced a reduction in pull-down efficiency upon longer storage.

8. The sonication steps may be unnecessary for some tissues, e.g., germ cells.

9. It is important to sonicate lightly. Too much sonication will lead to the formation of small debris that can stick to nuclei and beads and will decrease the purity of the final preparation.

10. The pellet can be examined under a microscope for the degree of fragmentation and presence of labeled nuclei.

11. We use an OptiPrep cushion to prevent nuclei from being forced against the wall of the conical tube, which would cause lysis and clumping.

12. It is important to use a swinging bucket rotor so that the nuclei are pelleted on top of the OptiPrep cushion and do not collect on the side of the tube.

13. We collect the nuclei by first removing the supernatant, then removing the OptiPrep through the layer of nuclei, to leave the nuclei in approximately 500–1,000 μL in the conical tube.

14. Washing the nuclei is necessary because of the abundance of endogenous biotinylated proteins in *C. elegans*, which compete with the biotinylated BLRP tag for binding of the streptavidin beads. As most of these proteins are cytoplasmic, background levels can be reduced by washing the nuclei. We found that a minimum of two washes are necessary.

15. It is possible to use beads conjugated to an anti-FLAG antibody to affinity-purify the nuclei via FLAG-tag without the need for in vivo biotinylation by BirA. We found that this reduced the purity of the affinity-purified nuclei from >90 to 80–90 %.

16. We tested several different sizes of beads and found that 2-μm beads gave the best results because of their size relative to the nuclei. Larger beads crushed the nuclei, whereas smaller beads were less efficient in capturing the nuclei.

17. Coating the 1-mL tip with BSA reduces sticking of the beads and nuclei to the tip, which increases the yield. BSA from some sources can contain RNases and should be avoided when the purified nuclei are used for expression profiling.

18. We found that two passes over the column give the best combination of yield and purity.

19. Bead-bound nuclei are considered positives, and non-bead-bound nuclei are considered false positives. The ratio of the number of non-bead-bound to bead-bound nuclei represents the purity of the pull-down, which should be >0.9. The purity can also be assessed by counting mCherry-positive and mCherry-negative nuclei. However, this approach is less practical, as there is strong auto-fluorescence of the magnetic beads. We have found that the two approaches result in very similar numbers. From the number of bead-bound nuclei and the size of the aliquot, the total number of nuclei in the pull-down sample (yield) can be extrapolated.

20. Milk should be avoided as a blocking agent, as it contains biotin that will cause background when probing with streptavidin-HRP.

21. Streptavidin detection tests for the successful in vivo biotinylation of the BLRP tag within the NTF. However, the signal is often relatively weak, possibly due to the presence of relatively large amounts of endogenous biotinylated proteins. We therefore also routinely detect the NTF with an anti-FLAG antibody, which also confirms the size of the NTF. We detect histone

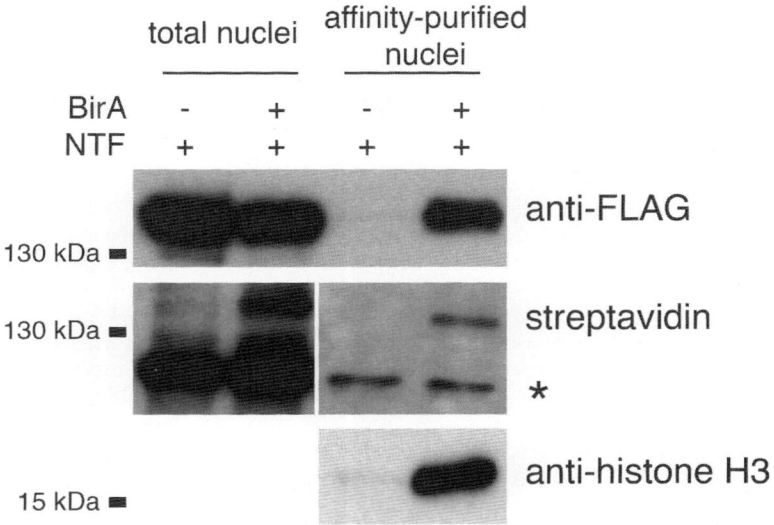

Fig. 3 Representative Western blots. Blots were probed by anti-FLAG, streptavidin, and anti-histone H3. The FLAG tag (*top*) is present in both input samples, but only in the successful pull-down, indicating the successful pull-down of the nuclear-tagging fusion (NTF) protein. Streptavidin (*middle*) confirms the in vivo biotinylation of the NTF. Endogenous biotinylated proteins are also visible and marked with an asterisk. Histone H3 (*bottom*) is only present in the successful pull-down, confirming that nuclei and not just the NTF were pulled down (image reproduced from [21])

H3 in the pull-down samples to confirm successful purification containing chromatin. Detection of histone H3 is also very sensitive to false positives in the negative control. Representative Western blots are shown in Fig. 3.

Acknowledgements

This work was supported by HHMI, NIH (U01-HG004274), and the Swiss National Science Foundation (PBSKP3-124362).

References

1. Ng RK, Gurdon JB (2008) Epigenetic inheritance of cell differentiation status. Cell Cycle 7:1173–1177

2. Yuan G, Zhu B (2012) Histone variants and epigenetic inheritance. Biochim Biophys Acta 1819:222–229

3. Azuara V, Perry P, Sauer S et al (2006) Chromatin signatures of pluripotent cell lines. Nat Cell Biol 8:532–538

4. Fox RM, Watson JD, Von Stetina SE et al (2007) The embryonic muscle transcriptome of Caenorhabditis elegans. Genome Biol 8:R188

5. Miller MR, Robinson KJ, Cleary MD, Doe CQ (2009) TU-tagging: cell type-specific RNA isolation from intact complex tissues. Nat Methods 6:439–441

6. Roy PJ, Stuart JM, Lund J, Kim SK (2002) Chromosomal clustering of muscle-expressed

genes in Caenorhabditis elegans. Nature 418: 975–979

7. Bonn S, Zinzen RP, Girardot C et al (2012) Tissue-specific analysis of chromatin state identifies temporal signatures of enhancer activity during embryonic development. Nat Genet 44:148–156

8. Bonn S, Zinzen RP, Perez-Gonzalez A et al (2012) Cell type-specific chromatin immunoprecipitation from multicellular complex samples using BiTS-ChIP. Nat Protoc 7:978–994

9. Fox RM, Von Stetina SE, Barlow SJ et al (2005) A gene expression fingerprint of C. elegans embryonic motor neurons. BMC Genomics 6:42

10. Sugiyama T, Rodriguez RT, McLean GW et al (2007) Conserved markers of fetal pancreatic epithelium permit prospective isolation of islet progenitor cells by FACS. Proc Natl Acad Sci U S A 104:175–180

11. Von Stetina SE, Watson JD, Fox RM et al (2007) Cell-specific microarray profiling experiments reveal a comprehensive picture of gene expression in the C. elegans nervous system. Genome Biol 8:R135

12. Haenni S, Ji Z, Hoque M et al (2012) Analysis of C. elegans intestinal gene expression and polyadenylation by fluorescence-activated nuclei sorting and 3'-end-seq. Nucleic Acids Res 40:6304–6318

13. Southall TD, Gold KS, Egger B et al (2013) Cell-type-specific profiling of gene expression and chromatin binding without cell isolation: assaying RNA Pol II occupancy in neural stem cells. Dev Cell 26:101–112

14. Burgemeister R (2011) Laser capture microdissection of FFPE tissue sections bridging the gap between microscopy and molecular analysis. Methods Mol Biol 724:105–115

15. Erickson HS, Albert PS, Gillespie JW, Rodriguez-Canales J, Marston Linehan W, Pinto PA et al (2009) Quantitative RT-PCR gene expression analysis of laser microdissected tissue samples. Nat Protoc 4: 902–922

16. Golubeva Y, Salcedo R, Mueller C et al (2013) Laser capture microdissection for protein and NanoString RNA analysis. Methods Mol Biol 931:213–257

17. Neira M, Azen E (2002) Gene discovery with laser capture microscopy. Methods Enzymol 356:282–289

18. Rabien A (2010) Laser microdissection. Meth Mol Biol 576:39–47

19. Deal RB, Henikoff S (2010) A simple method for gene expression and chromatin profiling of individual cell types within a tissue. Dev Cell 18:1030–1040

20. Deal RB, Henikoff S (2011) The INTACT method for cell type-specific gene expression and chromatin profiling in Arabidopsis thaliana. Nat Protoc 6:56–68

21. Steiner FA, Talbert PB, Kasinathan S et al (2012) Cell-type-specific nuclei purification from whole animals for genome-wide expression and chromatin profiling. Genome Res 22: 766–777

22. Amin NM, Greco TM, Kuchenbrod LM et al (2014) Proteomic profiling of cardiac tissue by isolation of nuclei tagged in specific cell types (INTACT). Development 141: 962–973

23. Henry GL, Davis FP, Picard S et al (2012) Cell type-specific genomics of Drosophila neurons. Nucleic Acids Res 40:9691–9704

Chapter 2

Lysis Gradient Centrifugation: A Flexible Method for the Isolation of Nuclei from Primary Cells

Karl Katholnig, Marko Poglitsch, Markus Hengstschläger, and Thomas Weichhart

Abstract

The isolation of nuclei from eukaryotic cells is essential for studying the composition and the dynamic changes of the nuclear proteome to gain insight into the mechanisms of gene expression and cell signalling. Primary cells are particularly challenging for standard nuclear isolation protocols due to low protein content, sample degradation, or nuclear clumping. Here, we describe a rapid and flexible protocol for the isolation of clean and intact nuclei, which results in the recovery of 90–95 % highly pure nuclei. The method, called lysis gradient centrifugation (LGC), is based on an iso-osmolar discontinuous iodixanol-based density gradient including a detergent-containing lysis layer. A single low g-force centrifugation step enables mild cell lysis and prevents extensive contact of the nuclei with the cytoplasmic environment. This fast method shows high reproducibility due to the relatively little cell manipulation required by the investigator. Further advantages are the low amount of starting material required, easy parallel processing of multiple samples, and isolation of nuclei and cytoplasm at the same time from the same sample.

Key words Nuclear isolation, Mild lysis, Fractionation, Single step, Discontinuous density gradient

1 Introduction

The spatial separation of the nuclear and the cytoplasmic compartment plays an essential role in the regulation of gene expression in all immune cells [1]. Gaining information about the composition and the dynamic changes in the nuclear proteome is important for understanding the mechanisms of gene regulation and cell signaling [2–4]. A number of protocols aim to isolate pure and intact nuclei from eukaryotic cells for further analysis by western blots, electrophoretic mobility shift assays (EMSA), or mass spectrometry [3–8]. The most important and initial step of frequently employed nuclear isolation protocols is the disruption of the cytoplasmic membrane, while the nuclear membrane should remain intact. This is usually achieved by induction of cell swelling through

Ronald Hancock (ed.), *The Nucleus*, Methods in Molecular Biology, vol. 1228,
DOI 10.1007/978-1-4939-1680-1_2, © Springer Science+Business Media New York 2015

the incubation of the cells in a hypotonic buffer for a tightly controlled period of time followed by mechanical or chemical disruption of the cell membrane [9]. NP-40 is a nonionic non-denaturing detergent and was for a long time the most frequent detergent used in nuclear isolation protocols. It was either directly included into the hypotonic lysis buffer or added after cell swelling [10, 11]. As NP-40 is no longer commercially available, it was replaced by the chemically indistinguishable IGEPAL CA-630. Mechanical stress is often used to aid plasma membrane disruption and is introduced either by defined vortexing times or by passing the cells through a syringe. The crude lysates containing the nuclei are then usually washed by pelleting the nuclei by centrifugation. Alternatively, Graham and coworkers developed a method to purify the nuclei from the crude lysate based on a discontinuous density gradient formed by iodixanol followed by centrifugation and nuclear banding [9]. This method has been used frequently to study nuclear proteins [12–14].

From a biochemical point of view, the purification of nuclei from mammalian cells is associated with many technical problems [11]. Temperature control cannot be assured easily at any time of the nuclear isolation procedure during mechanical disruption methods, and moreover the nuclei are in extensive contact with the cytoplasmic environment containing lysis-activated proteases that lead to partial degradation and modification of the samples despite the presence of protease inhibitors [15, 16]. Repeated washing steps, necessary to remove nucleus-associated membrane systems and cytoskeletal components, may result in nuclear leakage due to overexposure to detergent and mechanical stress. Therefore, nuclear leakage results not only in loss of target proteins but also in clumping of the nuclei due to released chromosomal DNA [17]. Clumping leads to a low recovery of nuclei, explaining the large amount of starting material required by standard fractionation procedures. Many protocols are optimized for distinct cell types as the nuclear density and stability show profound variations among cell types and stimulation procedures [18, 19]. All of these caveats are true for cell lines and primary cells; however, it is particularly difficult to get reliable and clean nuclear preparations from primary cells which are available only in a limited amount and have low protein content [20].

2   Materials

2.1   Cells

This method has been used successfully for:

Primary cells: CD14+ monocytes; macrophages (MΦ); T lymphocytes.

Cell lines: JE-6 (Jurkat cells, an immortalized line of T lymphocytes); RAJI (a lymphoblastoid cell line derived from a Burkitt lymphoma); MEF (murine embryonic fibroblasts); THP-1 cells (a promonocytic cell line derived from a human acute lymphocytic leukemia patient) (*see* **Note 1**).

2.2 Lysis Gradient Preparation

1. Visipaque™ 320 (65.2 % Iodixanol; GE Healthcare).

2. Phosphate-buffered saline solution (PBS): 210.0 mg/L KH$_2$PO$_4$, 9,000 mg/L NaCl, 726.0 mg/L Na$_2$HPO$_4$-7H$_2$O (Lonza).

3. Protease inhibitor cocktail: 10× solution made by dissolving one tablet of cOmplete EDTA-free (Roche) in 2 mL of PBS.

4. Coomassie Brilliant Blue R250 (Bio-Rad).

5. Crystal Violet (Sigma-Aldrich).

6. IGEPAL CA-630 (Sigma).

7. Polystyrene tubes: 8 mL, 13 × 100 mm (BD Biosciences).

8. Crushed ice.

2.3 Nuclear Gradient Centrifugation

1. Standard refrigerated laboratory centrifuge (e.g., Rotanta 460RS, Hettich) equipped with a swinging bucket rotor.

2. Vacuum-attached glass pipette.

2.4 Western Blots

1. Denaturing SDS-sample buffer (4×): 250 mM Tris–HCl, pH 6.8, 40 % glycerol, 8 % SDS, 400 mM dithiothreitol, and 0.04 % (w/v) bromophenol blue.

2. Nitrocellulose membranes: Protran, Whatman.

3. PBS-T: 0.05 % Tween-20 in PBS.

4. Blocking solution: 4 % dry milk (Bio-Rad) in PBS-T.

5. Primary antibodies: Rabbit anti-GAPDH mAb (Cell Signaling) 1:1,000 in PBS-T + 0.02 % NaN$_3$; rabbit anti-ABTF-IID polyclonal (Santa Cruz) 1:500 in PBS-T + 0.02 % NaN$_3$.

6. Secondary antibody: ECL donkey anti-rabbit IgG, HRP-linked whole Ab [GE Healthcare): 1:3,000 in PBS-T + 4 % dry milk.

7. HRP Substrate: Immobilon Western Chemiluminescent (Millipore).

8. Chemiluminescence imager.

3 Methods

Our method is based on an iodixanol-based gradient [21] where we included a lysis layer containing the detergent IGEPAL CA-630 [22]. The incorporation of a lysis layer allows the isolation of intact nuclei from living cells in a single centrifugation step, and therefore minimizes the exposure of the nuclei to cytoplasmic protease activity. The gradient for LGC is built in such a way that the whole suspension of living cells can be directly loaded onto the iodixanol-based

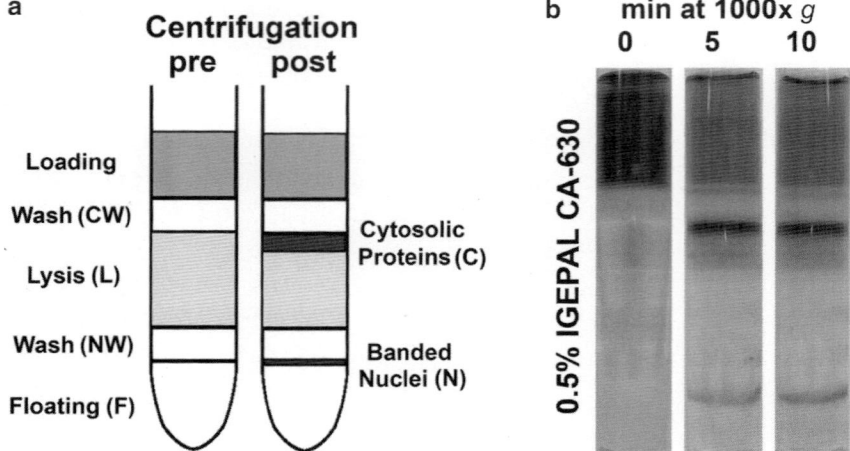

Fig. 1 (**a**) Principle of the method. The living cells are applied on top of the gradient in the original culture medium. During 10 min of low gravity centrifugation in a swinging bucket rotor, the cells pass through different functional layers including the cell wash layer (CW) and the lysis layer (L), at the beginning of which the cytoplasmic membrane is disrupted. Cytoplasmic proteins (C) remain on top of the lysis layer. Nuclei continue to pass through the nuclei wash layer (NW) and form a band on top of the floating layer (N). (**b**) A nuclear gradient purification performed with 2×10^6 THP-1 cells. The cells were centrifuged for the indicated times at $1,000 \times g$

gradient in its original culture medium (Fig. 1). The discontinuous iso-osmolar lysis gradient allows a g-force driven discrimination between unlysed and lysed cells and clean and ER-contaminated nuclei during centrifugation. A band of pure nuclei accumulates at the lowest interface, while nuclei contaminated with other cellular structures are trapped at upper layers due to their lower density (*see* **Note 2**). The cells can be stained with Crystal Violet and Coomassie Blue in order to visualize the bands occurring in the gradient and therefore enables easy harvesting of the nuclei.

3.1 Preparation of the Lysis Gradient

1. Prepare iodixanol dilutions up to 40 % (w/v) by diluting Visipaque 320 in PBS (*see* **Note 3**). The volume of Visipaque and PBS for the different layers of the gradient can be found in Table 1. Add one-tenth volume of 10× protease inhibitor solution to the cell lysis layer and nuclei wash layer immediately before use.

2. Add Coomassie Brilliant Blue to the lysis layer to a final concentration of 4 µg/mL (*see* **Note 4**).

3. Add Crystal Violet to the cell suspension and lysis layer to a final concentration of 5 µg/mL (*see* **Note 4**).

4. Add IGEPAL CA-630 to the lysis layer to a final concentration of 0.5 % (v/v).

5. Prepare lysis gradients in 8 mL polystyrene tubes suitable for 2×10^5–5×10^6 cells. The gradients are set up by sequentially overlaying 1 mL of floating layer with 0.5 mL of nuclei wash

Table 1
Composition of LGC gradients for isolating nuclei from different cell types

Iodixanol [% (w/v)]	Density [g/cm³]	PBS [mL]	Visipaque [mL]	Primary cells			Cell lines			
				CD14+	MΦ	T-Cells	THP-1	JE-6	RAJI	MEF
0	1.016	10.0	0.0							
5	1.051	9.2ᵃ	0.8	CW	CW	CW	CW	CW	CW	CW
10	1.071	8.5ᵃ	1.5	L	L	L	L	L	L	L
15	1.100	7.7	2.3							
20	1.129	6.9	3.1		NW					NW
25	1.155	6.2	3.8	NW		NW	NW	NW	NW	
30	1.178	5.4	4.6							
35	1.199	4.6	5.4	F	F	F	F	F	F	F
40	1.234	3.9	6.1							
65.2	1.372	0.0	10.0							

CW cell wash layer, *L* lysis layer, *NW* nucleus wash layer, *F* floating layer
ᵃTotal volume; includes 1 mL of 10× protease inhibitor solution added immediately before use

layer, 1 mL of lysis layer, and 0.5 mL of cell wash layer. All these steps are carried out on ice (*see* **Note 5**).

6. Stain the cooled cell suspension with 5 μg/mL Crystal Violet (*see* **Note 4**).

7. Apply the cooled cell suspension in up to 1 mL of the original culture medium immediately before lysis gradient centrifugation.

3.2 Nuclear Gradient Centrifugation and Harvesting of Fractions

1. After setting up the lysis gradient and adding the cell suspension, centrifuge at $1,000 \times g$ for 10 min at 4 °C in a refrigerated centrifuge equipped with a swinging bucket rotor.

2. After centrifugation, place the tubes carefully on ice.

3. Gently remove the culture medium as well as the upper part of the cell wash layer using a vacuum-attached glass pipette.

4. The upper 500 μL of the lysis layer are collected with a 1 mL Gilson pipette for the analysis of cytoplasmic fractions.

5. The residual lysis layer and the upper half of the nuclei wash layer are removed with a vacuum-attached glass pipette.

6. Collect the light blue band of nuclei (150 μL) between the nuclei wash layer and the floating layer (*see* **Note 6**).

7. For western blot analysis, lyse the nuclei directly by adding 50 µL of 4× reducing SDS-PAGE sample buffer followed by denaturation for 5 min at 95 °C (*see* **Note 7**).

3.3 Western Blots

1. Lyse the harvested fractions by adding one volume of 4× reducing SDS-PAGE sample buffer to three volumes of the fractions.

2. Denature proteins at 95 °C for 5 min.

3. Cool the samples to 10 °C and centrifuge in a microcentrifuge for 2 min at $25,000 \times g$.

4. Subject the supernatants to SDS-PAGE and blot the gels onto nitrocellulose membranes.

5. Incubate the membranes in blocking solution for 1 h at room temperature.

6. Incubate the membranes with primary antibodies at 4 °C overnight. In order to confirm the purity of the nuclear and cytoplasmic fractions, make use of antibodies against GAPDH and TF-IID: GAPDH should only be seen in the cytoplasmic fractions and the transcription factor TF-IID should only be present in the nuclear fraction (Fig. 2).

7. Wash the membranes 3× with PBS-T and add the horseradish peroxidase-labeled secondary antibody for 1 h at room temperature.

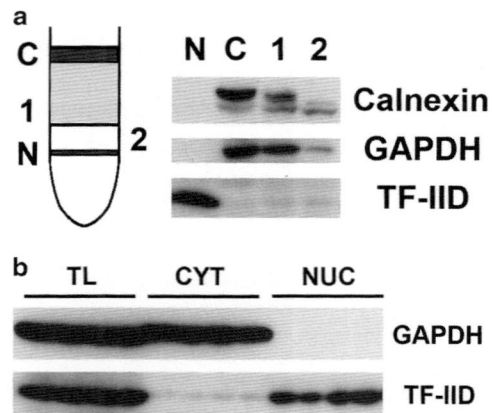

Fig. 2 Purity and integrity of LGC-isolated nuclei. (**a**) 10^6 THP-1 cells were subjected to LGC followed by taking aliquots of proteins for western blots from the cytoplasmic fraction (C), the lower interface of lysis layer (*1*), the central nuclear washing layer (*2*), and the nuclear band (N). Equal volumes of the indicated layers were western blotted and immunolabeled for Calnexin, GAPDH, and TF-IID. (**b**) Aliquots of nuclear (NUC) and cytoplasmic (CYT) fractions obtained by LGC from 10^6 THP-1 cells per treatment condition were investigated by western blotting for GAPDH and TF-IID. Total lysates (TL) in the same treatment conditions are shown as control

8. Wash the membranes 3× with PBS-T and detect antibody-labeled proteins with Immobilon Western Chemiluminescent HRP Substrate.

9. Detect the signals using a chemiluminescence imager.

4   Notes

1. This method can be used for the isolation of nuclei from virtually any cell type by just optimizing the densities of the different functional layers. It is also suitable for isolating nuclei from only a limited amount of cells; for example we have isolated nuclei from 5×10^5 primary macrophages and successfully performed western blotting. After an initial trypsinization step, adherent cells can also be processed with this protocol.

2. In principle, LGC can be adapted for ultracentrifugation to isolate cell organelles such as mitochondria, lysosomes, or peroxisomes requiring only a layer with the correct density to be present in the gradient [23, 24].

3. Visipaque 320 containing iodixanol is widely used as an intravenous contrast agent in clinical radiology, and the amount discarded from single-use containers can reach up to 50 % of the pack volume. The remaining Visipaque 320 provides a cheap and easily available source of iodixanol with the major advantage of iso-osmolarity, in contrast to OptiPrep which is a 60 % aqueous solution of Iodixanol whose iso-osmolarity after dilution cannot be controlled as precisely as with Visipaque [11].

4. If you want to carry out more sensitive methods than western blot after the LGC, it is recommended to avoid the use of Coomassie Brilliant Blue and Crystal Violet. The gradient can also be carried out without the addition of these dyes, which might interfere with more sensitive assays.

5. To minimize mixture of the layers when preparing the lysis gradient, the solutions should flow down the tube wall slowly and tubes should be placed on ice before the addition of the next layer. By omitting the lowest floating layer, the nuclei can be easily pelleted although this might influence their purity due to proteins sticking to the upper portion of the tubes.

6. Do not penetrate the wash layers with the pipette in order to harvest the nuclei, as proteins from the wash layer might stick to the pipette tip and contaminate the nuclei. It is highly recommended to suck off the medium and the upper half of the cell wash layer with a vacuum-attached glass pipette before harvesting the lysis layer for analysis of the cytoplasmic proteins. Subsequently, the residual layers and the upper half of the nuclei

wash layer should again be removed with the glass pipette before harvesting the nuclei.

7. The protein concentration of the banded nuclei collected in 150 μL is usually 0.25–0.5 mg/mL depending on the cell type. In order to measure the protein concentration or obtain a higher concentration in the sample the banded nuclei can be collected, diluted with 600 μL of PBS, pelleted by centrifugation at $1,000 \times g$, followed by protein extraction in a reduced volume of the desired extraction buffer.

Acknowledgements

We would like to thank Werner Poglitsch for technical support in initial experiments.

References

1. Graham JM, Rickwood D (1997) Subcellular fractionation. A practical approach. Oxford University Press, Oxford
2. Claude A (1975) The coming of age of the cell. Science 189:433–435
3. Barthelery M, Salli U, Vrana KE (2007) Nuclear proteomics and directed differentiation of embryonic stem cells. Stem Cells Dev 16:905–919
4. Albrethsen J, Knol JC, Jimenez CR (2009) Unravelling the nuclear matrix proteome. J Proteomics 72:71–81
5. Woodbury D, Len GW, Reynolds K et al (2008) Efficient method for gen erating nuclear fractions from marrow stromal cells. Cytotechnology 58:77–84
6. Cox B, Emili A (2006) Tissue subcellular fractionation and protein extraction for use in mass-spectrometry-based proteomics. Nat Protoc 1:1872–1878
7. Birnie GD (1978) Isolation of nuclei from animal cells in culture. Methods Cell Biol 17:13–26
8. Graham J, Ford T, Rickwood D (1994) The preparation of subcellular organelles from mouse liver in self-generated gradients of iodixanol. Anal Biochem 220:367–373
9. Blobel G, Potter VR (1966) Nuclei from rat liver: isolation method that combines purity with high yield. Science 154:1662–1665
10. Rio DC, Ares M Jr., Hannon GJ et al (2010) Preparation of cytoplasmic and nuclear rna from tissue culture cells. Cold Spring Harb Protoc pdb.prot5441
11. Graham JM (2001) Isolation of nuclei and nuclear membranes from animal tissues. Curr Protoc Cell Biol Chapter 3:Unit 3.10
12. Barta CA, Sachs-Barrable K, Feng F et al (2008) Effects of monoglycerides on P-glycoprotein: modulation of the activity and expression in Caco-2 cell monolayers. Mol Pharm 5:863–875
13. Valenzuela SM, Martin DK, Por SB et al (1997) Molecular cloning and expression of a chloride ion channel of cell nuclei. J Biol Chem 272:12575–12582
14. Zippin JH, Farrell J, Huron D et al (2004) Bicarbonate-responsive "soluble" adenylyl cyclase defines a nuclear cAMP microdomain. J Cell Biol 164:527–534
15. Hoffmann P, Chalkley R (1978) Procedures for minimizing protease activity during isolation of nuclei, chromatin, and the histones. Methods Cell Biol 17:1–12
16. Rickwood D, Messent A, Patel D (1997) Subcellular fractionation: a practical approach. IRL Press, Oxford, pp 71–105
17. Graham JM (2002) Rapid purification of nuclei from animal and plant tissues and cultured cells. ScientificWorld J 2:1551–1554
18. Dunphy WG, Rothman JE (1985) Compartmental organization of the Golgi stack. Cell 42:13–21
19. Goldberg DE, Kornfeld S (1983) Evidence for extensive subcellular organization of asparagin-linked oligosaccharide processing and lysosomal enzyme phosphorylation. J Biol Chem 258:3159–3165

20. Dyer RB, Herzog NK (1995) Isolation of intact nuclear extract preparation from a fragile B-lymphocyte cell line. Biotechniques 19: 192–195

21. Ford T, Graham J, Rickwood D (1994) Iodixanol: a nonionic iso-osmotic centrifugation medium for the formation of self-generated gradients. Anal Biochem 220:360–366

22. Poglitsch M, Katholnig K, Saemann MD et al (2011) Rapid isolation of nuclei from living immune cells by a single centrifugation through a multifunctional lysis gradient. J Immunol Meth 373:167–173

23. Cimini AM, Moreno S, Giorgi M, Serafini B, Ceru MP (1993) Purification of peroxisomal fraction from rat brain. Neurochem Int 23(3): 249–260

24. Graham JM (2001) Purification of a crude mitochondrial fraction by density-gradient centrifugation. Curr Protoc Cell Biol Chapter 3, Unit 3.4. doi:10.1002/0471143030. cb0304s04

Chapter 3

Isolation of Nuclei in Media Containing an Inert Polymer to Mimic the Crowded Cytoplasm

Ronald Hancock and Yasmina Hadj-Sahraoui

Abstract

Within cells, the nucleus is surrounded by the cytoplasm which contains diffusible macromolecules at a high concentration (>100 mg/ml). When cells are broken to isolate nuclei by current methods these macromolecules are dispersed, and to reproduce the environment of nuclei in vivo more closely we have developed a method to isolate them in a medium where cytoplasmic macromolecules are replaced by an inert, volume-occupying polymer and which is essentially cation-free. Nuclei isolated by this method resemble closely those prepared by conventional procedures as seen by optical and electron microscopy, and their internal compartments (nucleoli, PML and Cajal bodies, transcription centers, and splicing speckles) and transcriptional activity are conserved. This procedure is efficient for mammalian cells that normally grow in suspension and do not have an extensive cytoskeleton, and requires ~30 min.

Key words Isolation of nuclei, Nuclear compartments, Macromolecular crowding, Osmotic pressure, Ficoll, Dextran, Polyethylene glycol (PEG)

1 Introduction

Cell nuclei are commonly isolated in media containing cations at mM concentrations [1, 2] which screen the negative charges on chromatin and maintain its compaction [3, 4] and conserve nuclear ultrastructure and functions. Without these cations, isolated nuclei swell and may lyse due to the osmotic pressure caused by their high internal concentration of macromolecules [1]. However, standard isolation media may not reproduce the ionic environment in the cytoplasm which surrounds the nucleus in vivo, because in the cell K^+ and Na^+ ions are predominantly bound to macromolecules [5–8] and Mg^{2+} ions to ATP, macromolecules, mitochondria, and the sarcoplasmic reticulum [9, 10], and biophysical considerations suggest that "there is no cytoplasmic bulk concentration of ions and metabolites (as is often assumed in biophysical models and in the design and interpretation of in vitro experiments)" [11].

Ronald Hancock (ed.), *The Nucleus*, Methods in Molecular Biology, vol. 1228,
DOI 10.1007/978-1-4939-1680-1_3, © Springer Science+Business Media New York 2015

Significant quantities of extra Mg^{2+} and Ca^{2+} ions become bound to nuclei during their isolation in conventional media [12].

Since the cytoplasm contains diffusible macromolecules at a concentration in the range of 130 mg/ml [13–16] the nucleus, like other cytoplasmic organelles and macromolecular assemblies, is predicted to experience strong macromolecular crowding effects [17–19]. To mimic these crowding effects media containing proteins or inert polymers have been used to isolate mitochondria [20–22], peroxisomes [23], chloroplasts [24], Golgi apparatuses [25], and nucleoids of bacteria [26] but similar media have been employed only rarely to isolate nuclei [27, 28].

With these considerations in mind and recalling Arthur Kornberg's seventh commandment "Correct for extract dilution with molecular crowding" [29], we have developed a method to isolate nuclei in media in which an inert, volume-occupying polymer replaces the cytoplasmic macromolecules dispersed upon cell breakage and whose only ionic component is 100 µM K-Hepes buffer, pH 7.4. This environment is predicted to reproduce that of nuclei in vivo more closely than conventional isolation media. The polymers used are Ficoll or dextran of M_r 70 kDa, molecules of a size excluded from nuclei [30, 31] but which are inconvenient to handle and pipette due to the viscosity of their solutions, or polyethylene glycol (PEG) of M_r 8 kDa which penetrates through the nuclear envelope and exerts crowding effects on macromolecules and structures within the nucleus; PEG forms solutions of low viscosity which are convenient to use and provides nuclei with similar properties [31]. The cytoplasmic membrane is permeabilized by the non-ionic detergent digitonin, which releases cytoplasmic material while conserving the semipermeable properties of the nuclear membrane [32]. These nuclei are free of cytoplasmic material and closely resemble those prepared by conventional methods, as seen by optical and electron microscopy (Fig. 1), and their internal compartmentalization (Fig. 2) and transcriptional activity [31] are conserved, consistent with the idea that crowding by cytoplasmic macromolecules is an important but overlooked factor which determines the structure and functions of nuclei in vivo. This protocol is efficient for cell types that normally grow in suspension and do not have an extensive cytoskeleton, and requires ~30 min.

2   Materials

2.1   Cells

This method has been used successfully for cells of the lines Raji (human lymphoblastoid, ATCC CCL-86), K562 (human erythroleukemia, ATCC CCL-243), and P815 (mouse mastocytoma, ATCC TIB-64) and for human lymphocytes (*see* **Note 1**). Cells are harvested during exponential growth.

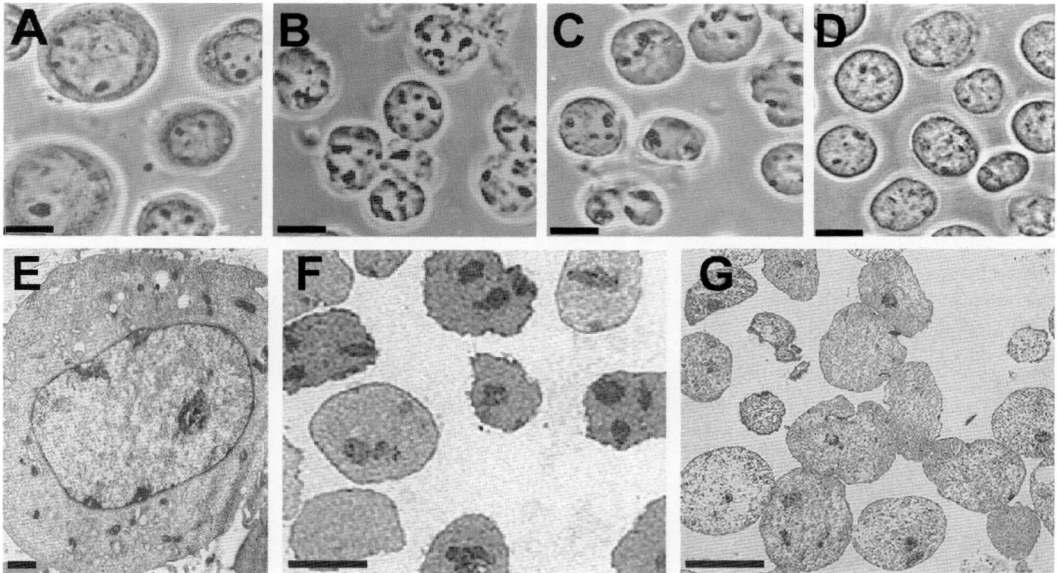

Fig. 1 Nuclei of K562 cells released in polymer solution containing 100 μM K-Hepes and digitonin. These nuclei are free of cytoplasmic material and closely resemble those prepared by conventional methods. (**a–d**) phase contrast images of (**a**) intact cells; (**b–d**) nuclei centrifuged onto slides after (**b**) homogenizing cells in 12 kDa PEG; (**c**) vortexing cells in 70 kDa Ficoll; (**d**) vortexing cells in 70 kDa dextran. (**e–g**) electron microscope sections of (**e**) intact cells; (**f**) nuclei isolated in PEG; (**g**) nuclei isolated in Ficoll. Sections were stained with uranyl acetate and lead citrate. Scale bars **a–d, f, g** = 10 μm, **e** = 1 μm. Modified from [31]

2.2 Reagents

1. 100 mM K-Hepes buffer (pH 7.4): to prepare 100 ml dissolve 2.38 g of Hepes in bidistilled H₂O. Adjust the solution to pH 7.4 by careful dropwise addition of 100 mM KOH while mixing on the pH meter, and filter through a 0.45 μm membrane filter. Store aliquots at −20 °C.

2. Polymer solutions: to prepare 50 ml of 12 % (w/v) 8 kDa PEG, 50 % (w/v) 70 kDa Ficoll, or 35 % (w/v) 70 kDa dextran, respectively, weigh 6 g of PEG (average M_r 8 kDa, Fluka molecular biology grade), 25 g of Ficoll (average M_r 70 kDa, Fluka), or 17.5 g of dextran (average M_r 69.9 kDa, Sigma-Aldrich) into a 50 ml conical polypropylene tube.

   Dissolve the polymer in bidistilled H₂O by incubating the solution in a thermostat at 60–70 °C with intermittent vortexing; PEG requires 1–2 h to dissolve while Ficoll and dextran may need several hours and vigorous vortexing. Cool the solution to room temperature and deionize by adding ~1 g of mixed-bed ion-exchange resin (AG 501-X8D, Bio-Rad) and mix occasionally over 6–8 h. Allow the resin beads to settle and pass the solution through an 0.45 μm syringe filter; dextran and Ficoll solutions filter slowly. Add 1/1,000 volume of 100 mM K-Hepes buffer (pH 7.4), verify the pH, and if necessary adjust to pH 7.4 by adding μl volumes of 100 mM

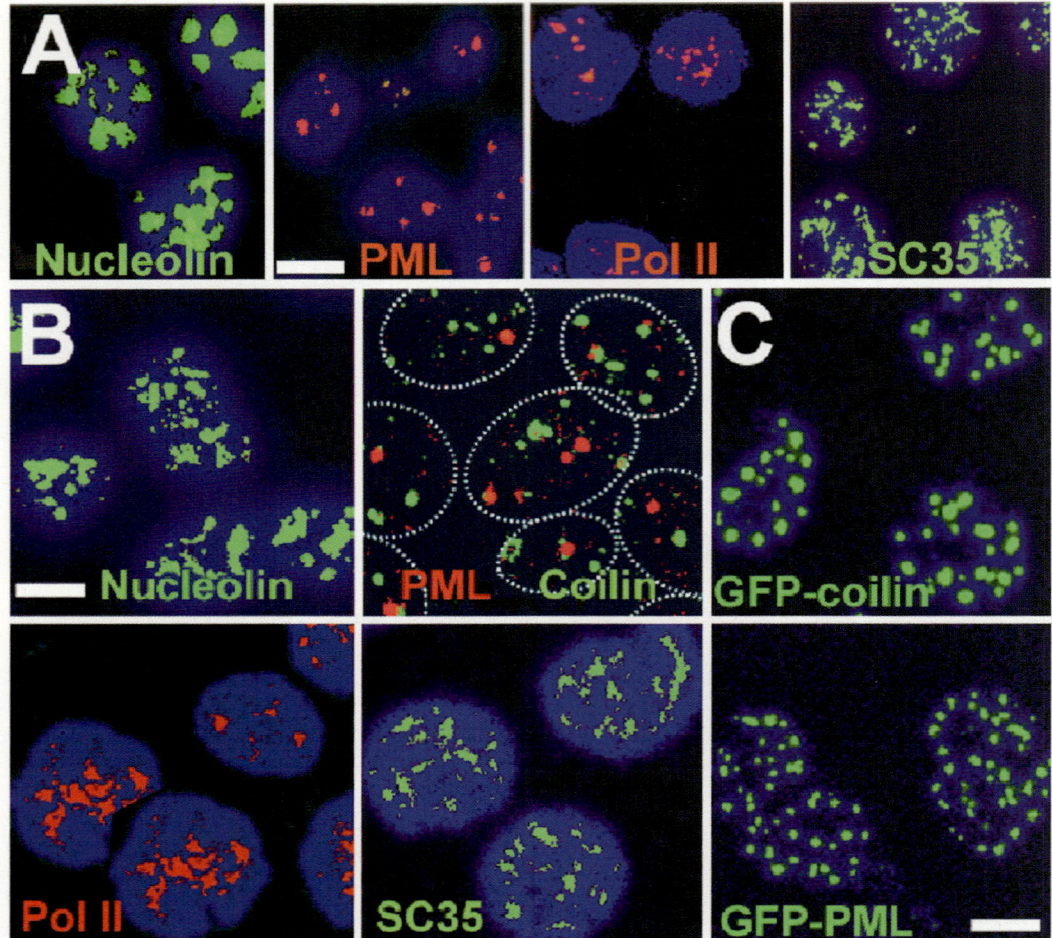

Fig. 2 Compartments are conserved in nuclei isolated in polymer solution. (**a**, **b**) Nuclei of K562 cells centrifuged onto slides and labeled with primary antibodies to the protein indicated followed by an appropriate secondary antibody conjugated with Alexa 488, 568, or 594, in order to visualize nucleoli, PML and Cajal bodies, transcription centers (RNA polymerase II), or splicing (SC35) speckles [33]. (**a**) Nuclei isolated in 12 kDa PEG or (**b**) in 70 kDa Ficoll. (**c**) Localization of coilin-GFP or PML isoform IV-GFP in nuclei of HeLa or U2OS cells, respectively, which express these proteins. Cells grown on slides were washed and permeabilized and extracted in situ with 70 kDa Ficoll solution containing digitonin (*see* **Note 1**). Images are maximum intensity projections from serial 0.5 μm confocal sections; DNA was stained with DAPI. The numbers of nucleoli, PML bodies, and Cajal bodies are essentially identical to those in nuclei isolated in a conventional ionic medium [31]. Scale bars = 5 μm. Modified from [31]

KOH while mixing on the pH meter. Store at 4 °C and verify the pH before each experiment using a plastic pH indicator strip (*see* **Note 2**).

3. Digitonin solution: dissolve digitonin in bidistilled H_2O at 10 mg/ml by incubation for 15 min at 98 °C; we use Boehringer high purity digitonin which is no longer available, but high-purity digitonin from other suppliers provides satisfactory

results. Cool the solution and store in aliquots at 4 °C; if a precipitate forms, redissolve by heating the solution in the same conditions.

4. Polymer solution supplemented with digitonin (100 µg/ml): prepare freshly for each experiment.

5. Two ml volume glass homogenizer with teflon piston (Potter-Elvehjem Type Tissue Grinder; Wheaton, Millville, NJ, USA).

6. Microscope slides coated with polylysine: these can be purchased, or alternatively dip slides in a 50 µg/ml solution of poly-L lysine for ~1 h, wash with deionized H_2O, drain well, and allow to dry in air.

3   Methods

Ficoll and dextran solutions at concentrations which maintain the volume of isolated nuclei are viscous and inconvenient to pipette. Solutions containing 12 % 8 kDa PEG are much less viscous and more convenient to use, and produce similar results. The buffering capacity of these solutions is low and their pH must be checked before each experiment. As described here, the method provides sufficient nuclei for imaging studies but it can be scaled up by increasing the volumes of solutions in proportion to the number of cells. All steps are carried out at room temperature, and nuclei are handled using micropipettes with ~5 mm cut from the tip.

1. To isolate nuclei in Ficoll or dextran, it is not practical to wash intact cells by centrifugation because of the high density and viscosity of these solutions. Instead, centrifuge ~10^7 cells a 15 ml conical polypropylene tube, remove all growth medium carefully with a micropipette and then with an absorbent paper wick, and resuspend the cells in 1 ml of polymer solution containing digitonin. After 10 min mix on a vortexer at maximum speed until >95 % of the cells have released their nucleus (usually 2–3 min); examine by phase-contrast microscopy after placing 5 µl samples on slides and covering with a cover glass. Add two volumes of the same polymer solution without digitonin to the suspension of nuclei and mix by inverting the tube.

2. To isolate nuclei in PEG solution, centrifuge ~10^7 cells in a 15 ml conical polypropylene tube, remove all the medium using a pipette attached to a vacuum line, and resuspend the cells in 1 ml of PEG solution by hand mixing. Centrifuge the cells ($600 \times g$, 10 min), resuspend by gentle hand mixing in 1 ml of PEG solution containing digitonin, and transfer the suspension to a 2 ml glass homogenizer with a teflon piston After 10 min, homogenize using slow hand strokes until >95 %

of the cells have released their nucleus, avoiding foaming (usually ~50 strokes); examine by phase-contrast microscopy (*see* Subheading 3, **step 1**). Add two volumes of the same PEG solution without digitonin and mix.

3. For optical (Fig. 1) or fluorescence imaging (Fig. 2) of nuclei, mount polylysine-coated slides in a cytological centrifuge, transfer aliquots of the suspension containing ~10^6 nuclei into the slide chambers, and centrifuge the nuclei onto the slides at $4,000 \times g$ for 40 min in Ficoll or dextran or $1,000 \times g$ for 10 min in PEG. For fixation, remove the solution by tapping the slides edge-wise on absorbent paper, overlay the nuclei with 0.5 ml of the same polymer solution used for the previous step supplemented with the desired fixative (we fix with 2 % (w/v) paraformaldehyde for 10 min). Incubate the slides in a humidified container (*see* **Note 3**).

4. If the nuclei must be washed for subsequent experiments, centrifuge them at $5,000 \times g$ for 15 min in Ficoll or dextran or $2,000 \times g$ for 5 min in PEG, resuspend them in the same polymer solution without digitonin, and centrifuge them in the same conditions (*see* **Notes 4–6**).

4   Notes

1. Fibroblastoid cells (e.g., HeLa, CHO, or U2OS cells) do not yield clean nuclei by this procedure because fibrous extracellular and cytoskeletal materials sediment with the nuclei and cannot be removed. Instead, these cells can be grown on slides or cover glasses and extracted in situ with the same solutions, leaving the nuclei attached to the surface. Remove all growth medium from slides or cover glasses carefully using a micropipette and absorbent paper wicks, overlay the cells with 500 μl of polymer solution for 5 min to wash them, and replace this solution with 500 μl of the same solution containing digitonin for 30 min to permeabilize and extract them (Fig. 2c).

2. PEG solutions may develop a slight yellow colour and their pH may decrease after storage for several weeks, due to unidentified processes, but this does not affect their efficacy in this method detectably.

3. A suitable humid chamber is a sealable plastic box containing a support to carry slides horizontally and moistened paper towels at the bottom.

4. If it is desired to examine the nuclei by electron microscopy, they can be centrifuged ($5,000 \times g$ for 15 min in Ficoll or dextran or $2,000 \times g$ for 5 min in PEG) and resuspended in the

same polymer solution containing fixative; we use 2 % (w/v) paraformaldehyde with 0.1 % (v/v) glutaraldehyde (both EM grade, Ted Pella, Redding, CA, USA) for 1 h on ice. The fixed nuclei can be centrifuged, embedded in 1.5 % (w/v) low melting-point agarose, dehydrated, embedded in an appropriate resin, sectioned, and stained with uranyl acetate and lead citrate (Fig. 1e–g) [31].

5. For further studies of isolated nuclei, for example of transcription or DNA replication, their osmotic conditions and morphology must be conserved by working in the same polymer solution used to isolate them.

6. Metaphase chromosomes can be isolated from mitotic cells in the same conditions [34].

Acknowledgements

We thank Jason Swedlow (Wellcome Trust Biocentre, University of Dundee, Scotland) for HeLa cells expressing GFP-coilin and David Bazett-Jones (Hospital for Sick Children, Toronto, Canada) for U2OS cells expressing GFP-PML isoform IV (originally from J. Taylor, Medical College of Wisconsin).

References

1. Anderson NG, Wilbur KM (1952) Studies on isolated cell components IV. The effect of various solutions on the isolated rat liver nucleus. J Gen Physiol 35:781–796

2. Busch H, Daskal Y (1977) Methods for isolation of nuclei and nucleoli. Methods Cell Biol 16:1–44

3. Aaronson RP, Woo E (1981) Organization in the cell nucleus: divalent cations modulate the distribution of condensed and diffuse chromatin. J Cell Biol 90:181–186

4. Engelhardt M (2004) Condensation of chromatin in situ by cation-dependent charge shielding and aggregation. Biochem Biophys Res Comm 324:1210–1214

5. Kellermayer M, Ludany A, Jobst K et al (1986) Cocompartmentation of proteins and K+ within the living cell. Proc Natl Acad Sci U S A 83:1011–1015

6. Edelmann L (1989) The physical state of potassium in frog skeletal muscle studied by ion-sensitive microelectrodes and by electron microscopy. Scanning Microsc 3:1219–1230

7. Ling GN (1990) The physical state of potassium ion in the living cell. Scanning Microsc 4:737–750

8. Negendank M, Shaller C (2005) Multiple fractions of sodium exchange in human lymphocytes. J Cell Physiol 104:443–459

9. Lüthi D, Günzel D, McGuigan JAS (1999) Mg-ATP binding, its modification by spermine, the relevance to cytosolic Mg2+ buffering, changes in the intracellular ionized Mg2+ concentration and the estimation of Mg2+ by 31P- NMR. Exp Physiol 84: 231–252

10. Günther T (2006) Concentration, compartmentation and metabolic function of intracellular free Mg2+. Magnes Res 19:225–236

11. Spitzer J, Poolman B (2005) Electrochemical structure of the crowded cytoplasm. Trends Biochem Sci 30:536–541

12. Naora H, Naora H, Mirsky AE et al (1961) Magnesium and calcium in isolated cell nuclei. J Gen Physiol 44:713–742

13. Arrio-Dupont M, Cribier S, Foucault G et al (1996) Diffusion of fluorescently labeled macromolecules in cultured muscle cells. Biophys J 70:2327–2332

14. Maughan DW, Godt RE (2001) Protein osmotic pressure and the state of water in frog myoplasm. Biophys J 80:435–442

15. Hou L, Lanni F, Luby-Phelps K (1990) Tracer diffusion in F-actin and Ficoll mixtures. Towards a model for cytoplasm. Biophys J 58: 31–43

16. Maughan DW, Henkin JA, Vigoreaux JO (2005) Concentrations of glycolytic enzymes and other cytosolic proteins in the diffusible fraction of a vertebrate muscle proteome. Mol Cell Proteomics 4:1541–1549

17. Cuneo P, Magri E, Verzola A et al (1992) 'Macromolecular crowding' is a primary factor in the organization of the cytoskeleton. Biochem J 281:507–512

18. Ellis RJ (2001) Macromolecular crowding: obvious but underappreciated. Trends Biochem Sci 26:597–604

19. Zhou HX, Rivas G, Minton AP (2008) Macromolecular crowding and confinement: biochemical, biophysical, and potential physiological consequences. Annu Rev Biophys 37:375–397

20. Birbeck MSC, Reid E (1956) Development of an improved medium for the isolation of liver mitochondria. J Biophys Biochem Cytol 2:609–624

21. Wicker U, Bücheler K, Gellerich FN et al (1993) Effect of macromolecules on the structure of the mitochondrial inter-membrane space and the regulation of hexokinase. Biochim Biophys Acta 1142:228–239

22. Laterveer FD, Gellerich FN, Nicolay K (2005) Macromolecules increase the channeling of ADP from externally associated hexokinase to the matrix of mitochondria. Eur J Biochem 232:569–577

23. Antonenkov VD, Sormunen RT, Hiltunen JK (2004) The behavior of peroxisomes in vitro: mammalian peroxisomes are osmotically sensitive particles. Am J Physiol Cell Physiol 287:C1623–C1635

24. Oku T, Kawahara H, Tomia G (1971) The Hill reaction and oxygen uptake in isolated pine chloroplasts. Plant Cell Physiol 12: 559–566

25. Morré DJ, Mollenhauer HH (1964) Isolation of the Golgi apparatus from plant cells. J Cell Biol 23:295–305

26. Cunha S, Woldringh CL, Odijk T (2001) Polymer-mediated compaction and internal dynamics of isolated Escherichia coli nucleoids. J Struct Biol 136:53–66

27. Takahashi Y, Asao T (1974) Study of the nuclei isolated from newt embryos by the use of Ficoll. Dev Growth Differ 16:281–294

28. Mason JA, Mellor J (1997) Isolation of nuclei for chromatin analysis in fission yeast. Nucleic Acids Res 25:4700–4701

29. Kornberg A (2000) Ten commandments, lessons from the enzymology of DNA replication. J Bacteriol 182:3613–3618

30. Seksek O, Biwersi J, Verkman AS (1997) Translational diffusion of macromolecule-sized solutes in cytoplasm and nucleus. J Cell Biol 138:131–142

31. Hancock R, Hadj-Sahraoui Y (2009) Isolation of cell nuclei using inert macromolecules to mimic the crowded cytoplasm. PLoS One 4:e7560

32. Cassany A, Gerace L (2009) Reconstitution of nuclear import in permeabilized cells. Methods Mol Biol 464:181–205

33. Handwerger KE, Gall JG (2006) Subnuclear organelles: new insights into form and function. Trends Cell Biol 16:19–26

34. Hancock R (2012) Structure of metaphase chromosomes: a role for effects of macromolecular crowding. PLoS One 7:e36045

Part II

Nucleoli

Chapter 4

A New Rapid Method for Isolating Nucleoli

Zhou Fang Li and Yun Wah Lam

Abstract

The nucleolus was one of the first subcellular organelles to be isolated from the cell. The advent of modern proteomic techniques has resulted in the identification of thousands of proteins in this organelle, and live cell imaging technology has allowed the study of the dynamics of these proteins. However, the limitations of current nucleolar isolation methods hinder the further exploration of this structure. In particular, these methods require the use of a large number of cells and tedious procedures. In this chapter we describe a new and improved nucleolar isolation method for cultured adherent cells. In this method cells are snap-frozen before direct sonication and centrifugation onto a sucrose cushion. The nucleoli can be obtained within a time as short as 20 min, and the high yield allows the use of less starting material. As a result, this method can capture rapid biochemical changes in nucleoli by freezing the cells at a precise time, hence faithfully reflecting the protein composition of nucleoli at the specified time point. This protocol will be useful for proteomic studies of dynamic events in the nucleolus and for better understanding of the biology of mammalian cells.

Key words Nucleoli, Rapid isolation, Protein composition, Quantitative proteomics

1 Introduction

The nucleolus has been a centre of active biological research for decades. Using isolated nucleoli, modern quantitative proteomics has identified more than 4,500 proteins in this organelle from different cell lines [1–3]. In addition to factors involved in ribosomal synthesis and assembly, nucleolar proteins are implicated in many important functions including RNA editing and mRNA maturation [4, 5], telomere metabolism [6, 7], viral infection response [8], microRNA functions [9] as well as stress responses [9, 10]. Quantitative fluorescent microscopy uncovers the dynamics of proteins in the nucleolus [11]. Most nucleolar proteins rapidly flux in and out of the nucleolus and their distribution between the nucleolus and the rest of the cell changes upon stimulation [2, 12]. For example, the expression level of some ribosomal proteins and chromosomal remodeling proteins changes rapidly within 10 min after serum stimulation [10].

Ronald Hancock (ed.), *The Nucleus*, Methods in Molecular Biology, vol. 1228,
DOI 10.1007/978-1-4939-1680-1_4, © Springer Science+Business Media New York 2015

Traditionally, nucleoli are isolated from cultured mammalian cells by using a two-step procedure in which nuclei are first isolated. Current methods are modifications of the protocol developed 50 years ago based on the principle that in the presence of a low concentration of Mg^{2+}, cell nuclei are broken by sonication while nucleoli remain intact [13]. These methods require pre-isolation of nuclei which includes detaching cells by trypsin (at 37 °C), washing them with PBS, swelling of the cells, dounce homogenization, and centrifugation onto a sucrose cushion which takes more than 1 h before we can proceed to nucleolar isolation, during which time some critical information on the nucleoli has already been missed or changed. A previous one-step protocol [1], although avoiding isolating nuclei and directly sonicate cells in the hypotonic buffer, fails to generate intact nucleoli, perhaps due to the hypotonic medium. To obtain intact nucleoli and capture rapid responses of nucleolar proteins requires a more sophisticated and faster method.

We have developed a new and improved nucleolar isolation method using adherent HeLa cells, in which the need for nuclear purification is bypassed. The cells are directly washed with a cold sucrose-containing buffer and harvested by scraping, then sonicated and centrifuged on a sucrose cushion (Fig. 1). The concentrations of sucrose and Mg^{2+} in the buffer are adjusted so that the cells are disrupted by sonication while nucleoli remain intact. The whole process can be finished in 20 min. Electron microscopy of sections of the isolated nucleoli indicates that they have similar

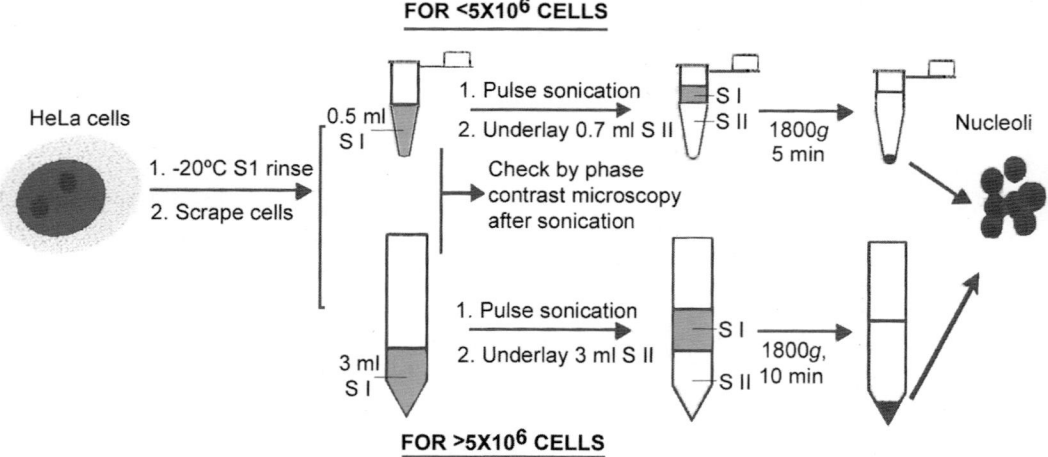

Fig. 1 Schematic diagram of this method for isolating nucleoli. Adherent HeLa cells are cooled rapidly in cold (−20 °C) sucrose solution (S I) and scraped from dishes on ice. For less than 5×10^6 cells, they are collected into an 1.5 mL eppendorf tube whereas if the number of cells is greater they are collected into 15 mL tubes. The cell suspension is underlaid with solution II (S II) and the tubes are centrifuged. Isolated nucleoli form a pellet. All steps are performed on ice or at 4 °C. Reprinted from [10] with permission from Elsevier

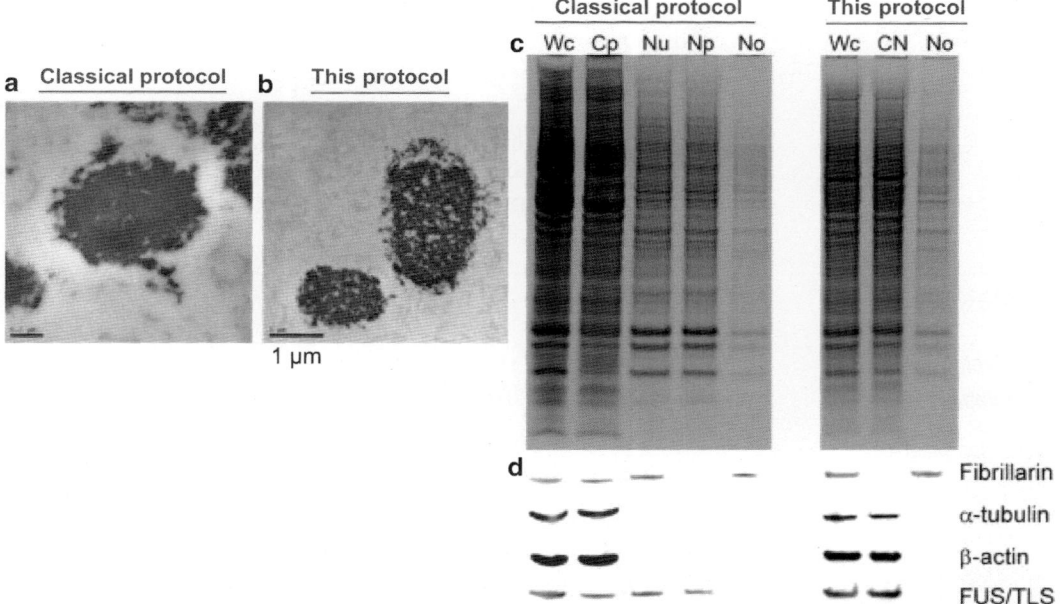

Fig. 2 Nucleoli isolated by this protocol compared with those prepared by a classical method [1]. (**a**, **b**) Sections of nucleoli by transmission electron microscopy. (**c**) Proteins in subcellular fractions separated by SDS-PAGE and stained with Coomassie Blue. Each lane was loaded with a sample derived from the same number of cells. *Wc* whole cells, *Cp* cytoplasm, *Nu* nuclei, *Np* nucleoplasm, *CN* cytoplasm + nucleoplasm, *No* nucleoli. (**d**) Detection of fibrillarin (a nucleolar marker), α-tubulin and β-actin (cytoplasmic markers), and FUS/TLS (a nucleoplasmic marker) in fractions by Western blotting. Adapted from [10] with permission from Elsevier

ultrastructure to those isolated by a classical method [13] (Fig. 2a), and they retain most of the cellular fibrillarin but are virtually devoid of cytoplasmic and nucleoplasmic marker proteins (Fig. 2). The majority of the proteins in nucleoli purified by this protocol and identified by SILAC labeling are also detected in nucleoli isolated by the classical method and function in ribosome biogenesis and other nucleolar functions [10]. Many proteins over-represented in nucleoli isolated by the classical method are related to mitochondria which are common contaminants, and the dense sucrose cushion in this method may help to remove mitochondria from the nucleoli. Their yield and purity are consistent across experiments and relatively insensitive to changes in the starting cell number. There are several advantages to this method: (a), it can start with very few cells ($\sim 5 \times 10^5$); (b), it is a simple and fast 4-step procedure consisting of washing-scraping-sonication-centrifugation (Fig. 1); (c), the purity of the nucleoli is good (Fig. 2a) and the yield is high; and (d), the first step of incubating cells in cold (−20 °C) sucrose solution freezes their biochemical status and is therefore able to capture any changes at precisely controlled time points and to characterize biochemical events in the nucleolus within less than 10 or even 5 min after a particular treatment.

Some ribosomal proteins such as RL31 and proteins involved in chromosome remodeling were found to respond very rapidly (within 10 min) to serum-starvation and -replenishing; notably, the proliferation marker Ki-67 enters the nucleolus 5 min after serum replenishment [10]. To our knowledge, this is the first study that demonstrates such fast responses in the nucleolus, further confirming the rapid plasticity of this organelle.

2 Materials

2.1 Isolation of Nucleoli

2.1.1 Consumables and Equipment

1. Culture medium: DMEM, fetal calf serum, penicillin-streptomycin solution (all from Invitrogen).

2. Culture dishes: 10 cm-diameter dishes, petri dishes, or 6-well plates depending on the amount of nucleoli to be isolated (*see* **Note 1**).

3. Cell scraper (*see* **Note 2**).

4. Centrifuge tubes: 1.5 or 15 mL plastic.

5. Sonicator: VCX130 (Sonics & Materials, Newtown, USA) (*see* **Note 3**).

6. Centrifuge: swinging bucket type capable of spinning 15 mL tubes at $1,800 \times g$.

7. Phase contrast microscope (*see* **Note 4**), glass slides, and coverslips.

2.1.2 Reagents and Solutions

1. Complete protease inhibitor tablets (Roche).

2. 2.5 M sucrose stock solution: 855.75 g sucrose/L in autoclaved distilled H_2O (*see* **Note 3**).

3. 2 M $MgCl_2$ stock solution: 9.52 g $MgCl_2$/50 mL in distilled H_2O.

4. Solution I (S I): 0.5 M sucrose, 3 mM $MgCl_2$ with one tablet of protease inhibitors/50 mL, store at 4 °C (*see* **Note 4**).

5. Solution II (S II): 1.0 M sucrose, 3 mM $MgCl_2$ with one tablet of protease inhibitors/50 mL, store at 4 °C.

2.2 Sodium Dodecyl Sulfate Polyacrylamide Gel Electrophoresis (SDS-PAGE)

1. RIPA buffer: 50 mM Tris–HCl, pH 8, 150 mM NaCl, 0.1 % SDS, 0.5 % Na deoxycholate, 1 % Triton X-100, 1 mM PMSF, one protease inhibitor tablet/50 mL.

2. Dithiothreitol (DTT) (GE Healthcare) stock solution: 50 mM in H_2O, store in aliquots in freezer.

3. Urea: 8 M solution (24 g urea in 25 mL of H_2O).

4. Iodoacetamide (GE Healthcare) stock solution: 100 mM in H_2O, store in aliquots in freezer.

5. SDS-PAGE sample loading buffer (2×): 62.5 mM Tris–HCl, pH 6.8, 25 % (v/v) glycerol, 2 % (w/v) SDS, 0.01 % (w/v) bromophenol blue.

6. 10 % Bis-Tris SDS-PAGE gels (NuPAGE, Life Technologies).

7. Running buffer (NuPAGE, Life Technologies).

8. Gel stain (SimplyBlue, Life Technologies).

2.3 Western Blotting

1. Primary antibodies: rabbit anti-human fibrillarin (Santa Cruz), mouse monoclonal anti-human (Sigma) β-actin (Sigma), mouse monoclonal anti-human α2-tubulin [Sigma), and mouse anti-human FUS/TLS (Santa Cruz).

2. Secondary antibodies: horseradish peroxidase (HRP)-labeled goat anti-mouse HRP and goat anti-rabbit HRP (both Santa Cruz).

3. Phosphate buffered saline (PBS): 3.2 mM Na_2HPO_4, 0.5 mM KH_2PO_4, 1.3 mM KCl, 135 mM NaCl.

4. Skimmed milk solution: 5 % (w/v) skimmed milk powder in PBS.

5. PBS–Tween-20 (PBST): 0.05 % (v/v) Tween-20 in PBS.

6. Enhanced chemiluminescence assay: HRP-based (ECL Plus, Amersham.

7. Chemiluminescence imager: Fuji LAS 4000 or similar.

3 Methods

3.1 Isolation of Nucleoli

3.1.1 Freezing and Collecting Cells

1. Seed HeLa cells in 10 cm petri dishes in DMEM supplemented with 10 % fetal calf serum, 100 μg/mL streptomycin and 100 u/mL penicillin and culture at 37 °C in 5 % CO_2 until >90 % confluence. One hour before nucleolar isolation, replace medium with fresh, pre-warmed medium (*see* **Note 5**).

2. Precool solution I at –20 °C. Decant the cell culture medium, quickly add 30 mL of cold solution I to the dish, and rinse twice with this solution.

3. Decant the solution I quickly, place the dish on ice, and scrape the cells from the dish (*see* **Note 6**).

4. Quickly transfer the cells into a 1.5 mL eppendorf tube. Add cold (–20 °C) solution I to a total volume of 0.5 mL (*see* **Note 7**).

3.1.2 Sonication and Sucrose Gradient Centrifugation

1. Sonicate the cells on ice at 50 % power, 10 s on, 10 s off, for five cycles (*see* **Note 8**).

2. Underlay the sonicated cell suspension with 0.7 mL of precooled solution II (*see* **Note 9**).

3. Centrifuge the tube at $1,800 \times g$ for 5 min at 4 °C (*see* **Note 10**). The resulting pellet contains the isolated nucleoli.

4. Discard the supernatant carefully without disturbing the sucrose layers (*see* **Note 11**). Transfer the pellet of nucleoli to a new tube.

5. Assess the yield and purity of the isolated nucleoli by phase contrast microscopy.

6. (Optional) To obtain isolated nucleoli of higher purity, resuspend the pellet in 0.5 mL of Solution I and repeat **steps 2–4** (*see* **Note 12**).

3.1.3 Analysis of Proteins

1. Add 10 mM DTT in 8 M urea to samples of whole cells, the cytoplasm + nucleoplasm fraction, and the isolated nucleoli and incubate at 56 °C for 40 min, followed by addition of iodoacetamide to 50 mM and incubation at room temperature in the dark for 40 min.

2. Mix the samples with an equal volume of 2× gel loading buffer and incubate for 10 min at room temperature.

3. Load an aliquot of each cell fraction from the same number of cells (5×10^5) on a 10 % SDS-PAGE gel and electrophorese at 200 V for 35 min.

4. Stain the gel with colloidal Coomassie Blue (Fig. 2c).

3.1.4 Western Blot

1. Transfer proteins from the gel to a nylon membrane at 100 V for 1 h.

2. Block the membrane in 5 % skimmed milk for 1 h at room temperature, wash twice in PBST for 15 min.

8. Incubate with primary antibodies for 1 h (anti-fibrillarin 1:1,000; anti-tubulin 1:1,000; anti-actin 1:4,000; anti-FUS/TLS 1:1,000).

3. Wash the membrane twice with PBST for 15 min.

4. Incubate the membrane with goat anti-mouse HRP or goat anti-rabbit HRP (both Santa Cruz, 1:2,000) for 1 h, wash twice with PBST for 15 min.

5. Detect the signals using a chemiluminescence imager (Figs. 2d and 3).

4   Notes

1. Here we describe the isolation of nucleoli from 2×10^7 cells for proteomics studies, but this method also allows isolation of nucleoli from a very small number of cells, for example from one well of a 6-well plate.

2. Please quickly scrap the cells on ice.

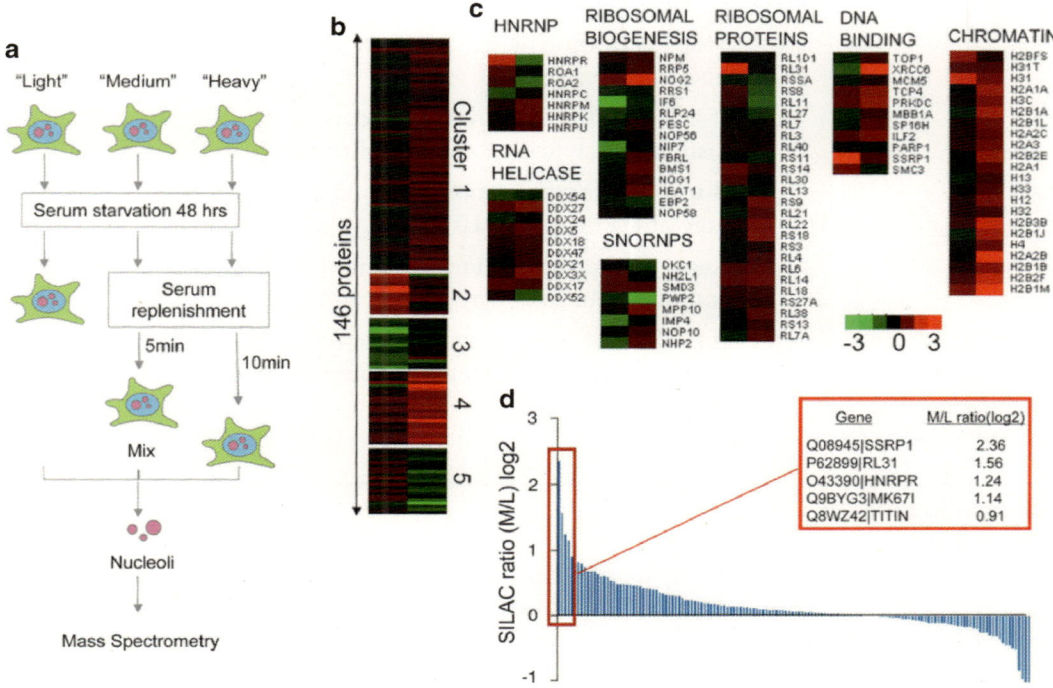

Fig. 3 An example of the application of the method described here to study rapid changes in the nucleolar proteome of HeLa cells in response to serum stimulation using SILAC proteomic methods. (**a**) Cells were cultured in serum-free medium containing different heavy isotope-labeled amino acids (termed "light," "medium," and "heavy") for 48 h, and replenished with serum for 0, 5, or 10 min. The cells were collected and mixed together. Nucleoli were isolated and subjected to mass spectrometry analysis. (**b**) Proteome changes within 5 and 10 min of serum replenishment, clustered in terms of their dynamics. (**c**) Groups that belong to specific cellular components used as examples to show the dynamic changes of each protein group. (**d**) Changes in nucleolar proteins after 5 min of serum replenishment. The *box* highlights the five proteins with the highest accumulation. Reprinted from [10] with permission from Elsevier

3. Do not autoclave the sucrose solutions. Bacterial and other organism will not grow in saturated sucrose solution.

4. Add protease inhibitors for proteomics studies, whereas add RNase inhibitors for analysis of RNA in the isolated nucleoli.

5. This step keeps the cells in optimal condition, but is optional.

6. For proteomic studies, use a cell scraper rather than trypsin to collect cells because trypsin digestion at 37 °C may change the nucleolar proteome [10].

7. If collecting cells from more than one dish for preparing a large amount of nucleoli, perform the sucrose centrifuge step in a 15 mL tube.

8. It is important to ensure that the probe of the sonicator is immersed in the solution. Optimize the sonication time when applying this protocol to other cell lines by checking the cells

by phase contrast microscopy after each sonication pulse. Stop sonication when 90 % of the cells are broken and the nucleoli are released.

9. If using 3 mL of solution II to collect cells from more than one dish, underlay with 3 mL of solution I in a 15 mL tube.

10. If using cells from more than one dish in 15 mL tubes, centrifuge at $1{,}800 \times g$ for 10 min.

11. Remove the supernatant as completely as possible, because it contains abundant cytoplasmic proteins.

12. If nucleoli of very high purity are required, we suggest repeating the sucrose cushion step but some nucleoli may be lost during this additional step.

Acknowledgements

This project was funded by a General Research Fund (project number 9041520) provided by the Research Grant Council, Hong Kong. We thank Longman Liang, Sarah Cheung, Myra Cheung, and Leo Sodium in Dr. Lam's lab for their assistance in developing this new method.

References

1. Andersen JS, Lyon CE, Fox AH et al (2002) Directed proteomic analysis of the human nucleolus. Curr Biol 12:1–11

2. Lam YW, Lamond AI, Mann M et al (2007) Analysis of nucleolar protein dynamics reveals the nuclear degradation of ribosomal proteins. Curr Biol 17:749–760

3. Ahmad Y, Boisvert FM, Gregor P et al (2009) NOPdb, nucleolar proteome database – 2008 update. Nucleic Acids Res 37:D181–D184

4. Jellbauer S, Jansen RP (2008) A putative function of the nucleolus in the assembly or maturation of specialized messenger ribonucleoprotein complexes. RNA Biol 5:225–229

5. Sansam CL, Wells KS, Emeson RB (2003) Modulation of RNA editing by functional nucleolar sequestration of ADAR2. Proc Natl Acad Sci U S A 100:14018–14023

6. Kieffer-Kwon P, Martianov I, Davidson I (2004) Cell-specific nucleolar localization of TBP-related factor 2. Mol Biol Cell 15:4356–4368

7. Zhang S, Hemmerich P, Grosse F (2004) Nucleolar localization of the human telomeric repeat binding factor 2 (TRF2). J Cell Sci 117:3935–3945

8. Wang L, Ren XM, Xing JJ, Zheng AC (2010) The nucleolus and viral infection. Virol Sin 25:151–157

9. Li ZF, Liang YM, Lau PN et al (2013) Dynamic localisation of mature microRNAs in human nucleoli is influenced by exogenous genetic materials. PLoS One 8:e70869

10. Liang YM, Wang X, Ramalingam R et al (2012) Novel nucleolar isolation method reveals rapid response of human nucleolar proteomes to serum stimulation. J Proteomics 77:521–530

11. Dundr M, Hebert MD, Karpova TS et al (2004) In vivo kinetics of Cajal body components. J Cell Biol 164:831–842

12. Trinkle-Mulcahy L, Lamond AI (2007) Toward a high-resolution view of nuclear dynamics. Science 318:1402–1407

13. Busch H, Muramatsu M, Adams H et al (1963) Isolation of nucleoli. Exp Cell Res 24(Suppl 9):150–163

Chapter 5

Sequential Recovery of Macromolecular Components of the Nucleolus

Baoyan Bai and Marikki Laiho

Abstract

The nucleolus is involved in a number of cellular processes of importance to cell physiology and pathology, including cell stress responses and malignancies. Studies of macromolecular composition of the nucleolus depend critically on the efficient extraction and accurate quantification of all macromolecular components (e.g., DNA, RNA, and protein). We have developed a TRIzol-based method that efficiently and simultaneously isolates these three macromolecular constituents from the same sample of purified nucleoli. The recovered and solubilized protein can be accurately quantified by the bicinchoninic acid assay and assessed by polyacrylamide gel electrophoresis or by mass spectrometry. We have successfully applied this approach to extract and quantify the responses of all three macromolecular components in nucleoli after drug treatments of HeLa cells, and conducted RNA-Seq analysis of the nucleolar RNA.

Key words Nucleolus, DNA, Protein isolation, RNA, TRIzol

1 Introduction

The nucleolus is a nuclear structural component involved in ribosome production, RNA processing, and the regulation of mitosis, cell growth, and programmed cell death [1–3]. Cellular stresses induce major morphological changes in the nucleolus, and quantitative data suggest these changes are associated with reorganization of the nucleolar proteome [4, 5]. However, the study of these proteomic effects in relation to the complex macromolecular context of the nucleolus requires efficient extraction and accurate quantification of both RNA and proteins from the purified nucleoli.

Nucleoli can be efficiently separated from other cellular compartments, but most buffers used for protein extraction, e.g., RIPA (Tris–HCl buffer containing 1 % NP-40 and 1 % SDS) do not perform well with nucleolar preparations [6]. While lysis of the nucleoli into denaturing gel loading buffer yields high recovery of protein [7, 8], this is not compatible with protein quantification

Ronald Hancock (ed.), *The Nucleus*, Methods in Molecular Biology, vol. 1228,
DOI 10.1007/978-1-4939-1680-1_5, © Springer Science+Business Media New York 2015

assays. Sonication, DNase treatment and high salt buffers have not yielded substantial improvements in the extraction efficiency of nucleolar proteins [6, 9]. The majority of studies of the nucleolus have focused on only one macromolecular component (protein, RNA, or DNA), and no single method has been applicable to simultaneous study of the changes in all three macromolecular components from the same nucleolar specimen.

We have developed a TRIzol-based method that efficiently and sequentially releases RNA, DNA, and protein components from the nucleolus and that can be used simultaneously for isolation of cytoplasmic and nuclear compartments. TRIzol is a robust lysis and denaturing reagent, which is primarily designed for RNA extraction and has generally not been used for the extraction of DNA and protein. While buffers for protein extraction are generally designed for isolation of soluble proteins, the TRIzol reagent lyses entire organelles [10] and precipitates both soluble and insoluble proteins. Until now, TRIzol-extracted proteins have been difficult to dissolve using the solubilizing reagents recommended by the producer of TRIzol (Invitrogen) [11]. However, applying a specially designed buffer (SDS–urea–Tris buffer, SUBT) this limitation is now readily overcome [12]. Comparing the TRIzol based method to lysis of whole cells by urea–Tris buffer (UBT) or by the PARIS kit cell disruption buffer (Ambion) showed nearly identical recoveries of proteins over a wide size range [12]. As applied to sequential RNA, DNA, and protein extraction from the cytoplasmic, nuclear, and nucleolar fractions of HeLa cells, the method demonstrated ability to yield high molecular weight DNA while at the same time allowing dynamic studies of specific RNA and protein components in these three cellular compartments [12, 13]. Solubilized proteins were readily quantified by standard methods, and the quantities of isolated nucleolar RNA and protein were highly correlated [12]. Individual proteins varying over a size range from 20 to 200 kDa were readily identified by specific antibodies, allowing studies of nucleolar protein dynamics in the presence of an RNA polymerase I or nuclear export inhibitor [12, 13]. Thus, given the small sample volumes and the difficulties of extracting proteins by standard methods, we expect the TRIzol method to be of particular value for studies of molecular dynamics in the nucleolus.

2   Materials

All solutions should be prepared in ultrapure deionized water, and all chemicals should be of analytical grade. Appropriate handling procedures and safety protocols should be adhered to when handling the biological materials and potentially hazardous reagents. Institutional policies should be strictly adhered to in this regard.

2.1 Cell Culture and Growth

1. Appropriate culture medium for the cell type to be used. For example, HeLa (human cervical adenocarcinoma) cells are cultured in Dulbecco's Modified Eagle's Medium (DMEM, Invitrogen) supplemented with 10 % fetal calf serum.

2. Phosphate-buffered saline solution without calcium and magnesium (PBS): 210.0 mg/L KH_2PO_4, 9,000 mg/L NaCl, 726.0 mg/L $Na_2HPO_4 \cdot 7H_2O$.

3. Solution of 0.05 % (w/v) trypsin and ethylenediamine tetraacetic acid (Trypsin-EDTA, Invitrogen).

4. Fifteen centimeters diameter cell culture-grade polystyrene dishes.

2.2 Reagents

1. TRIzol (Invitrogen).

2. Isopropanol.

3. Ethanol.

4. Chloroform (CH_3Cl).

5. DEPC-treated deionized water.

6. 1 M Tris–HCl, pH 7.4.

7. Urea powder.

8. Sodium citrate powder.

9. Guanidine hydrochloride powder.

10. 2.5 M sucrose stock solution in H_2O.

11. 1 M $MgCl_2$ solution in H_2O.

12. 5 M NaCl solution in H_2O.

13. Dithiothreitol (DTT): 0.5 M solution in H_2O.

14. IGEPAL CA-630 (Sigma).

15. BCA protein assay kit (Pierce).

16. Sonicator equipped with a 6.4 mm (1/4 in.) microtip: Misonix (Farmingdale, NY, USA) or similar.

17. Microvolume UV–Vis spectrophotometer: NanoDrop, Quibit, or equivalent.

2.3 Buffers

The following solutions should be freshly prepared, in DEPC-treated deionized water unless stated otherwise:

1. Nuclear extraction buffer: 50 mM Tris–HCl, pH 7.4, 0.14 M NaCl, 1.5 mM $MgCl_2$, 0.5 % IGEPAL CA-630 (a nonionic, non-denaturing detergent), 1 mM DTT.

2. S1 solution: 0.25 M sucrose, 10 mM $MgCl_2$.

3. S2 solution: 0.35 M sucrose, 0.5 mM $MgCl_2$.

4. S3 solution: 0.88 M sucrose, 0.5 mM $MgCl_2$.

5. Sodium citrate–ethanol solution: 0.1 M sodium citrate in 10 % ethanol.

6. Guanidine hydrochloride 0.3 M in 95 % ethanol.

7. SUTB buffer: mix equal volumes of 1 % SDS solution and UTB buffer (75 mM Tris–HCl pH 7.5, 9 M urea, and 0.15 M β-mercaptoethanol).

3   Methods

3.1   Isolation of Nucleoli

3.1.1   Preparation of HeLa Cells

1. Plate HeLa cells in 15 cm-diameter cell culture dishes.

2. Once the cells reach 90 % confluence, rinse them once with PBS and detach them with 2.5 mL of trypsin/EDTA solution.

3. Inactivate the trypsin by addition of 2.5 mL of growth medium, and collect the cells in a 50 mL centrifuge tube.

4. Count the cells. Usually we collect 1.8×10^8 cells.

5. Centrifuge the cells for 5 min at $100 \times g$.

6. Wash the cell pellet with 20 mL cold PBS and centrifuge for 5 min at $300 \times g$ at 4 °C. Repeat two additional times.

3.1.2   Purification of Nuclei

All steps should be carried out on ice or at 4 °C.

1. Resuspend the cells by adding 10 mL of nucleus extraction buffer [14] with gentle pipetting, making sure there are no clumps of cells.

2. Incubate the cell suspension on ice for 5 min. During the last minute, spot 2 μL of cell suspension on a slide and check under the microscope. Normally, at least 90 % of the cells will have released their nuclei (*see* **Note 1**).

3. Separate the nuclear and cytoplasmic fractions by centrifugation for 5 min at $300 \times g$ at 4 °C.

4. Remove the supernatant (cytoplasmic fraction). A 200 μL aliquot of the supernatant may be used to check the efficiency of separation (*see* **Note 2**).

5. Resuspend the pellet (nuclear fraction) in 3 mL of buffer S1.

6. Prepare a fresh 50 mL centrifuge tube with 3 mL of buffer S2 and gently layer the nuclear suspension on top of the buffer S2.

7. Centrifuge for 5 min at $2,000 \times g$ at 4 °C.

8. Resuspend the purified nuclear pellet in 3 mL of buffer S2.

9. Wash the sonicator tip several times with deionized H_2O and then with 70 % ethanol, and dry.

10. Sonicate the nuclear suspension on ice using power setting 1. Sonicate for 10 s, then wait for 20 s. Spot 2 μL of the sonicated suspension on a slide and check under the phase contrast

microscope. Free nucleoli should be readily observed as dense, refractive bodies.

11. Continue sonication until all nuclei are disrupted (*see* **Note 3**).

12. Carefully layer the sonicated nuclear suspension on top of 3 mL of buffer S3 in 15 mL centrifuge tube.

13. Collect the nucleoli by centrifugation for 10 min at $3,000 \times g$ at 4 °C.

14. Resuspend the nucleolar pellet in 500 µL buffer S2 and centrifuge for 5 min at $1,430 \times g$ at 4 °C.

15. Nucleoli can be stored in buffer S2 at −80 °C for a prolonged period, or lysed immediately (*see* Subheading 3.2).

16. Optional: stain the purified nucleoli with antibodies to nucleolar proteins, for example NPM, FBL, or UBF, to confirm their integrity and purity.

3.2 Sequential Purification of RNA, DNA and Protein Using the TRIzol Reagent

3.2.1 Extraction of RNA from Nucleoli

1. Resuspend the purified nucleoli in 1 mL of TRIzol reagent. Since the nucleolar pellet/precipitate solubilizes slowly, vortexing the suspension normally improves solubility (*see* **Note 4**).

2. Incubate the lysate for an additional 5 min at room temperature.

3. Add 200 µL of cold chloroform and mix by vortexing.

4. Incubate the mixture for 3 min.

5. Centrifuge the mixture for 15 min at $12,000 \times g$ at 4 °C.

6. Carefully transfer the aqueous phase of the sample (around 600 µL), which contains the RNA components, to a 1.5 mL microtube. Save the volume containing the interphase and phenol–chloroform phase for DNA isolation and subsequent protein extraction (*see* Subheadings 3.2.2 and 3.2.3) (*see* **Note 5**).

7. To the aqueous phase (the RNA-containing fraction), add 500 µL of 100 % isopropanol (at room temperature) to precipitate the RNA.

8. Mix the sample by inverting the tube several times and incubate at room temperature for 10 min.

9. Centrifuge for 10 min at $12,000 \times g$ at 4 °C. If no RNA pellet is visible, *see* **Note 6**.

10. Prepare 80 % ethanol and cool on ice.

11. Carefully remove the supernatant from the tube. Do not disturb the RNA pellet.

12. Wash the pellet with 1 mL of ice-cold 80 % ethanol. Invert the tube to resuspend the RNA pellet.

13. Centrifuge the tube for 5 min at $12,000 \times g$ at 4 °C.

14. Remove the supernatant and air-dry the RNA pellet for 10 min.

15. Dissolve the RNA pellet in 50 μL of DEPC-treated deionized water.

16. Quantify the RNA sample using a NanoDrop or equivalent instrument. The quality is acceptable when the 260/280-ratio is around 2.0 and the 260/230-ratio is around 2.1.

17. Store the RNA at –80 °C.

18. Optional: Visualize RNA integrity by agarose gel electrophoresis.

*3.2.2 Extraction of DNA from the Interphase and the Phenol–Chloroform Phase (See Subheading 3.2.1, **Step 6**)*

1. Remove the remaining aqueous phase overlying the interphase.

2. Add 300 μL 100 % ethanol per 1 mL of TRIzol used for the initial lysis (*see* Subheading 3.2.1, **step 1**).

3. Cap the tube and mix by inverting the tube several times.

4. Incubate the sample for 2–3 min at room temperature.

5. Centrifuge the tube for 5 min at $2,000 \times g$ at 4 °C to pellet DNA.

6. Remove the phenol–ethanol supernatant and save it for protein isolation (*see* Subheading 3.2.3).

7. Wash the DNA pellet with 1 mL of sodium citrate–ethanol solution per 1 mL of TRIzol reagent (*see* Subheading 3.2.1, **step 1**).

8. Incubate for 30 min at room temperature with constant rotation.

9. Collect the DNA pellet by centrifugation for 5 min at $2,000 \times g$ at 4 °C. Discard the wash supernatant. Repeat the wash step one more time.

10. Add 2 mL 75 % of ethanol per 1 mL of TRIzol reagent (*see* **Note 7**).

11. Incubate the sample for 30 min at room temperature with constant rotation.

12. Centrifuge for 5 min at $2,000 \times g$ at 4 °C, and discard the supernatant.

13. Air-dry the DNA pellet for 5–10 min. Do not over-dry.

14. Dissolve the DNA pellet in 50 μL of water. Incubation in a 50 °C water bath may improve the solubility.

15. Remove any insoluble material by centrifuging the sample at $12,000 \times g$ for 10 min.

16. Transfer the supernatant containing the DNA to a new tube.

17. Measure the DNA concentration using a NanoDrop or equivalent instrument.

18. Optional: Visualize DNA integrity on a 0.8 % TAE agarose gel.

*3.2.3   Extraction of Proteins from the Phenol–Ethanol Phase (See Subheading 3.2.2, **Step 6**)*

1. Divide the phenol–ethanol phase (approximately 900 μL) (*see* Subheading 3.2.2, **step 6**) into two equal aliquots and transfer to fresh 1.7 mL microtubes.

2. Add 750 μL of isopropanol to each of the approximately 450 μL aliquots of the phenol–ethanol phase.

3. Incubate for 10 min at room temperature. A white protein precipitate can be observed during this step.

4. Centrifuge for 10 min at $12,000 \times g$ at 4 °C to pellet the proteins.

5. Discard the supernatant and wash the protein pellet in each tube with 1 mL of 0.3 M guanidine hydrochloride in 95 % ethanol.

6. Incubate the tubes for 20 min at room temperature with constant mixing (*see* **Note 8**).

7. Centrifuge for 5 min at $7,500 \times g$ at 4 °C. Discard the supernatant wash solution. Repeat the washing step two more times.

8. Wash the protein pellet in each tube with 1 mL of 100 % ethanol.

9. Incubate for 20 min at room temperature with constant rotation.

10. Centrifuge for 5 min at $7,500 \times g$ at 4 °C. Discard the supernatant wash solution.

11. Air-dry the protein pellet for 10 min.

12. Add 100 μL of SUTB buffer to the protein pellet and mix by micropipetting up and down (*see* **Note 9**).

13. Centrifuge the protein solution for 10 min at $12,000 \times g$ at room temperature to remove any insoluble material.

14. Transfer the supernatant containing the proteins to a new tube.

15. Dilute the protein sample 1:5 or 1:10 and measure the protein concentration using the Pierce BCA protein assay following the manufacturer's instructions.

16. Optional: Visualize the proteins by SDS-PAGE followed by Coomassie Brilliant Blue staining.

4   Notes

1. Incubation for 5 min is sufficient for the cell lines we have tested (HeLa, HCT116, WS1). A longer incubation time may be required for other cell lines.

2. The cytoplasmic fraction can be separated on an SDS-PAGE gel followed by western blotting analysis of nuclear or cytoplasmic proteins, e.g., lamin A or GAPDH, respectively, to verify purity.

3. A high sonication power can release the nucleoli quickly, but the yield of nucleoli is very low. Overall, the power provided by different sonicators varies and the settings should be tested empirically.

4. Stop vortexing when the solution appears homogenous. This ensures maximal recovery of nucleolar components.

5. Isolation of DNA and protein extraction should preferably be carried out within 24 h.

6. If no RNA pellet is visible after centrifugation, RNA-free glycogen can be added to 50–150 µg/mL to improve precipitation of RNA.

7. At this step, the DNA sample can be stored in 75 % ethanol at 4 °C for several months.

8. The protein pellet can be stored in the wash solution at −20 °C for several months.

9. Protein pellet isolated from the cytoplasmic compartment normally dissolve readily. For protein pellet isolated from the nuclear and nucleolar compartments, heating the sample at 55 °C will significantly improve the solubility. The protein pellet is usually completely dissolved in less than 1 h.

Acknowledgements

The original work leading to the development of this protocol has been supported by NIH P30 CA006973 and by Johns Hopkins University start-up funds.

References

1. Shaw PJ, Highett MI, Beven AF et al (1995) The nucleolar architecture of polymerase I transcription and processing. EMBO J 14:2896–906

2. Boisvert FM, van Koningsbruggen S, Navascues J et al (2007) The multifunctional nucleolus. Nat Rev Mol Cell Biol 8:574–585

3. Pederson T (1998) The plurifunctional nucleolus. Nucleic Acis Res 26:3871–3876

4. Andersen JS, Lam YW, Leung AK et al (2005) Nucleolar proteome dynamics. Nature 433:77–83

5. Moore HM, Bai B, Boisvert FM et al (2011) Quantitative proteomics and dynamic imaging of the nucleolus reveal distinct responses to UV and ionizing radiation. Mol Cell Proteomics 10(M111):009241

6. Chamousset D, Mamane S, Boisvert FM et al (2010) Efficient extraction of nucleolar proteins for interactome analyses. Proteomics 10:3045–3050

7. Scherl A, Coute Y, Deon C et al (2002) Functional proteomic analysis of human nucleolus. Mol Biol Cell 13:4100–4109

8. Andersen JS, Lyon CE, Fox AH et al (2002) Directed proteomic analysis of the human nucleolus. Curr Biol 12:1–11

9. Dignam JD, Lebovitz RM, Roeder RG (1983) Accurate transcription initiation by RNA polymerase II in a soluble extract from isolated mammalian nuclei. Nucleic Acids Res 11:1475–1489

10. Chomczynski P, Sacchi N (1987) Single-step method of RNA isolation by acid guanidinium

thiocyanate-phenol-chloroform extraction. Anal Biochem 162:156–159

11. Hummon AB, Lim SR, Difilippantonio MJ et al (2007) Isolation and solubilization of proteins after TRIzol extraction of RNA and DNA from patient material following prolonged storage. Biotechniques 42:467–470

12. Bai B, Laiho M (2012) Efficient sequential recovery of nucleolar macromolecular components. Proteomics 12:3044–3048

13. Bai B, Moore HM, Laiho M (2013) CRM1 and its ribosome export adaptor NMD3 localize to the nucleolus and affect rRNA synthesis. Nucleus 4:315–325

14. Liao JY, Ma LM, Guo Y et al (2010) Deep sequencing of human nuclear and cytoplasmic small RNAs reveals an unexpectedly complex subcellular distribution of miRNAs and tRNA 3' trailers. PLoS One 5:e10563

Part III

Genes and Chromatin

Chapter 6

Au Nanoinjectors for Electrotriggered Gene Delivery into the Cell Nucleus

Mijeong Kang and Bongsoo Kim

Abstract

Intracellular delivery of exogenous materials is an essential technique required for many fundamental biological researches and medical treatments. As our understanding of cell structure and function has been improved and diverse therapeutic agents with a subcellular site of action have been continuously developed, there is a demand to enhance the performance of delivering devices. Ideal intracellular delivery devices should convey various kinds of exogenous materials without deteriorating cell viability regardless of cell type and, furthermore, precisely control the location and the timing of delivery as well as the amount of delivered materials for advanced researches.

In this chapter the development of a new intracellular delivery device, a nanoinjector made of a Au (gold) nanowire (a Au nanoinjector) is described in which delivery is triggered by external application of an electric pulse. As a model study, a gene was delivered directly into the nucleus of a neuroblastoma cell, and successful delivery without cell damage was confirmed by the expression of the delivered gene. The insertion of a Au nanoinjector directly into a cell can be generally applied to any kind of cell, and a high degree of surface modification of Au allows attachment of diverse materials such as proteins, small molecules, or nanoparticles as well as genes on Au nanoinjectors. This expands their applicability, and it is expected that they will provide important information on the effects of delivered exogenous materials and consequently contribute to the development of related therapeutic or clinical technologies.

Key words Nanoinjector, Au nanowire, Nanoelectrode, Gene delivery, Intranuclear delivery

1 Introduction

Technologies for delivering exogenous materials into cells play a central role in fundamental research on cellular and molecular biology and diverse medical applications such as drug development. An ideal delivery device should convey various types of exogenous materials into cells without cellular damage regardless of the cell type and, more sophisticatedly, enable precise localization and timing of delivery as well as quantitative molecular dosing for advanced researches.

Ronald Hancock (ed.), *The Nucleus*, Methods in Molecular Biology, vol. 1228,
DOI 10.1007/978-1-4939-1680-1_6, © Springer Science+Business Media New York 2015

Recently, one-dimensional nanomaterials have been widely used in various studies due to their ability to directly access specific intracellular regions with minimum cell damage [1–3]. Nanoinjectors consisting of a carbon nanotube [4] or boron nitride nanotubes [5, 6] delivered quantum dots into single cells for the investigation of biological processes and biophysical properties of the cell. It was also shown that an atomic force microscopy tip etched to a needle-like structure with submicron diameter could inject genes into a single cell [7]. These techniques could be alternative methods in cases where conventional delivery methods targeting an entire cell population such as gene delivery into neurons [8] or primary culture cells [9] do not work effectively.

Increasing the cellular compatibility of delivery devices and maximizing the efficacy of the delivered materials are important for subsequent experiments following intracellular delivery. This can be accomplished by improving the material properties of nanoinjectors and the delivery method. Here, a new nanoinjector made of a gold nanowire (a Au nanoinjector) is described that has been developed for electrotriggered delivery of exogenous materials into a single cell with high spatial precision and quantitative dosage [10]. A Au nanowire can be inserted deep into a cell without significant cellular damage since it has a very thin cylindrical structure with an extremely small diameter (~100 nm) that is uniform along its whole length (a few tens of micrometers) [11]. In addition, the superb mechanical flexibility of a Au nanowire originating from its perfectly crystalline nature allows it to pass the heterogeneous interior of a cell compliantly, which further increases cellular compatibility by reducing cell deformation during the insertion of the nanoinjector [11].

The efficacy of intracellularly delivered materials is critically influenced by the location of the delivery [12]. The extremely thin Au nanoinjector can provide maximum efficacy by delivering materials to an appropriate subcellular site of action with high spatial resolution. Timing of delivery is also important, especially for precise studies concerning a specific cell cycle phase or cellular state [13]. The high electrical conductivity of a Au nanowire enables electrotriggered delivery employing the external application of an electrical pulse to release the materials attached to the nanowire surface, which can regulate the timing of delivery. By controlling the parameters of electric pulses and, in addition, adjusting the attachment of materials on the Au nanowire, quantitative delivery can be achieved. Particularly, the atomically smooth and well-defined surface of a Au nanowire provides an ideal canvas for the formation of a self-assembled monolayer (SAM), which facilitates the control of attachment of versatile materials on the nanowire via SAM [14, 15].

In this Chapter, Au nanoinjector-based electrotriggered delivery of a linear DNA fragment and a plasmid into the nucleus of neuroblastoma cells is described as a model study and successful delivery is demonstrated by expression of the delivered gene. This delivery system, however, does not limit the types of either cells or delivered materials. Readers should optimize the detailed experimental conditions related to their cells, the material to deliver, the linker molecule to anchor the material to the Au nanoinjector, and the parameters of the electrical pulses for releasing the material when considering the appropriate delivery for their experimental system of choice.

2 Materials

2.1 Au Nanoinjectors

2.1.1 Preparing Au Nanowires

1. Horizontal furnace system: horizontal hot-wall single-zone furnace with an inner quartz tube (1 in. diameter) with a vacuum pump and temperature, pressure and flow controllers.

2. A Au (gold) slug (a roughly shaped piece before processing).

3. Sapphire substrate (Hi-Solar, Korea).

4. Argon (Ar) gas.

2.1.2 Making Au Nanowire Electrodes

1. Optical microscope with CCD camera.

2. Manipulator: three-axis microstage with an attached home-built stand for holding a tungsten tip.

3. Tungsten tips (GGB Industries, Naples, FL, USA).

4. Conducting paste: a mixture of Ag paste (Norland Products, Cranbury, NJ, USA) and Ag nanopowder (Sigma-Aldrich).

5. Insulating materials: nail varnish and UV-curable polymer (Norland Products).

2.1.3 Making Au Nanoinjectors

1. DNA solution: Dissolve DNA in 1 M KH_2PO_4 aqueous solution, pH 6.7.

2. Cysteamine: 20 mM aqueous solution.

3. Rinsing solution: 0.2 % (w/v) aqueous sodium dodecyl sulfate.

2.1.4 Electrical Delivery System

1. Electrochemical workstation (CH Instruments, Austin, TX, USA).

2. Counter electrode: Pt wire (Alfa Aesar, Ward Hill, MA, USA).

3. Reference electrode: mercury sulfate electrode (ALS, Tokyo, Japan).

2.2 DNA

2.2.1 Dye-Labeled DNA for Fluorescence Observation

1. 23S ribosomal DNA (rDNA) from *Neisseria gonorrhoeae* (ATCC 10150).

2. PCR synthesis of thiolated rDNA: Taq buffer, dGTP, dATP, dTTP, dCTP, dCTP-Cy3 (GE Healthcare), Taq polymerase, Ngo13Fw forward primer (5′-HS-$(CH_2)_6$ GCGAAGTAGAAT AACGACGCATC-3′), MS38R reverse primer (5′-CCCGA CAAGGAATTTCGCTACCTTA-3′).

2.2.2 DNA Quantification	1. Reagents for PCR: *see* Subheading 2.2.1, **item 2**.
	2. TRIzol-LS reagent for DNA extraction (Life Technologies).
	3. Primers for quantitative polymerase chain reaction (qPCR): Ngo13Fw forward primer, Ngo3R reverse primer (5′-TTACCTACCCGTTGACTAAGTAAGC-3′).

2.2.3 DNA Encoding Enhanced Green Fluorescent Protein (EGFP)

1. EGFP-coding plasmid with CMV promoter and SV40 polyadenylation signals at 5′ and 3′ regions of the gene.

2. Synthesis of thiolated linear EGFP-coding DNA: thiol-modified forward primer 5′-HS-(CH2)6-TTACCGGAT AAGGCGCAGCG-3′ with the reverse primer (5′-CGCCCTTT GACGTTGGAGTC-3′).

3. DNA-spin plasmid DNA extraction kit (iNtRON Biotechnology, Gyeonggi-do, Korea).

2.3   Cells

2.3.1   Cultivation

1. SK-N-SH cells: human brain epithelial cells (ATCC HTB-11)

2. Culture medium: Dulbecco's Modified Eagle Medium (DMEM) with fetal bovine serum (10 % v/v), penicillin (100 units/mL, streptomycin (0.1 mg/mL); cells maintained in a 5 % CO_2, water-saturated atmosphere at 37 °C.

3. Gelatin-coated glass slides: slides treated with 0.01 % (w/v) gelatin solution (Sigma-Aldrich) for 30 min.

2.3.2   Testing Cell Viability

Trypan blue stain: 0.4 % (w/v) solution in phosphate-buffered saline (PBS) (210.0 mg/L KH_2PO_4, 9 g/L NaCl, 726.0 mg/L $Na_2HPO_4.7H_2O$).

---

3   Methods

3.1   Fabrication of Au Nanoinjectors with Immobilized DNA

Single-crystalline Au nanowires are grown via a vapor transport method [16]. A Au slug is evaporated at 1,100 °C in a horizontal furnace system. The Au vapor is carried to a c-cut sapphire substrate placed a few centimeters downstream by the flow of Ar gas at a rate of 100 sccm (standard cubic centimeters/minute) under a chamber pressure of ~10 Torr. Au nanowires grow vertically on the substrate from the half-octahedral Au seed naturally formed on the sapphire substrate. A single Au nanowire is attached to a macroscopic conducting support, a tungsten tip, for facile manipulation. The tungsten part is then insulated to avoid unnecessary electrochemical reactions occurring on the tungsten. The resulting Au nanowire–tungsten tip is called a Au nanowire electrode. After immobilizing DNA onto the surface of a Au nanowire electrode, it is called a Au nanoinjector.

3.1.1 Synthesis of Au Nanowires

1. Cut c-cut sapphire substrates to ~5×5 mm and place them in a low-temperature downstream zone in a horizontal quartz tube ~5 cm from the center of the heating zone.

2. Place a Au slug in an alumina boat at the center of the heating zone.

3. Flow Ar gas at a rate of 100 sccm with the chamber pressure maintained at 5–15 Torr.

4. Heat the Au slug at 1,100 °C for 30 min to 1 h.

5. Cool the substrate down to room temperature, maintaining the flow of Ar.

3.1.2 Fabrication of Au Nanowire Electrodes

1. Place a three-axis microstage near an optical microscope and fix the stand for holding a tungsten tip on the microstage.

2. Mount a tungsten tip on the stand.

3. Place the sapphire substrate with grown Au nanowires vertically to the stage of the optical microscope as shown in Fig. 1 (*see* **Note 1**).

4. Under optical monitoring, manipulate a tungsten tip to approach a Au nanowire and pick up the nanowire by softly touching it, as shown in Fig. 1 (*see* **Notes 2–6**).

5. Immerse the middle of the tungsten tip in a drop of nail varnish to insulate the tungsten part of the combined Au nanowire–tungsten tip, and move it back and forth up to the end of the tip (*see* **Notes 7** and **8**). Take out the nanowire–tungsten tip from the solution and dry it in air. This results in a Au nanowire electrode.

3.1.3 Immobilization of DNA on Au Nanoinjectors

1. Immerse a Au nanowire electrode in a solution of thiolated linear DNA (100 nM in 1 M KH_2PO_4) (*see* Subheading 2.2.3, **item 2**) at room temperature for 9 h (*see* **Note 9**). Covalent Au-S bonds are formed spontaneously between the nanowire and the thiolated DNA.

2. To immobilize plasmid DNA on a Au nanowire electrode, first immerse the electrode in an aqueous solution of cysteamine

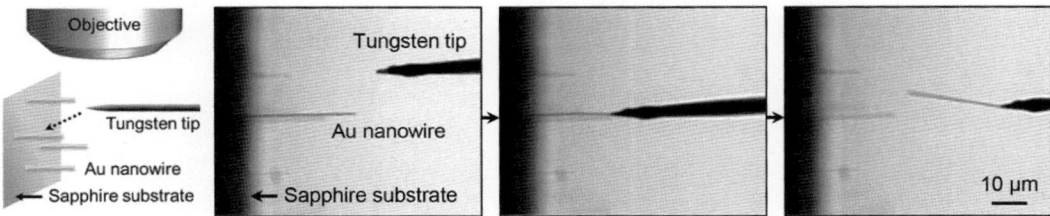

Fig. 1 Schematic illustration and optical images showing the process of attachment of a Au nanowire to a tungsten tip (Adapted with permission from [10]. Copyright 2013 American Chemical Society)

(20 mM) at room temperature for 30 min (*see* **Note 9**). Au-S covalent bonds are formed spontaneously between the Au nanowire and the cysteamine. After taking the Au nanowire electrode out from the cysteamine solution, immerse it in 1 M aqueous KH_2PO_4 solution and then take it out to remove excess cysteamine. Repeat this three times. Second, immerse this Au nanowire electrode in a 100 nM plasmid DNA solution in 1 M KH_2PO_4 at room temperature for 9 h. Electrostatic attraction occurs between the negatively charged plasmid DNA and the Au nanowire, which is positively charged in a neutral pH solution due to the immobilized cysteamine.

3. After taking out the Au nanowire electrode from the DNA solution, immerse it in SDS solution for 5 min to remove excess DNA (*see* **Note 9**). After taking out the Au nanowire electrode from the SDS solution, immerse it in a 1 M KH_2PO_4 solution and then take it out to remove remaining SDS. Repeat this three times. This results in a Au nanoinjector.

3.2 Electrochemical Release of DNA from a Au Nanoinjector

Dye-labeled DNA is immobilized on a Au nanoinjector, which is then immersed in an electrolyte solution in which counter and reference electrodes are also immersed. A voltage pulse with a selected magnitude and duration is applied to the Au nanoinjector to release the DNA. Then, the remaining labeled DNA on the nanoinjector is measured by imaging it in a fluorescence microscope.

3.2.1 Immobilization of Dye-Labeled DNA on a Au Nanoinjector

Linear DNA labeled with Cy3 for fluorescence observation is prepared by PCR amplification of 23S ribosomal DNA from *Neisseria gonorrhoeae* (ATCC 10150).

1. Perform PCR in 50 μL reactions containing 1× Taq buffer, 0.2 mM dGTP, 0.2 mM dATP, 0.2 mM dTTP, 0.125 mM dCTP, 0.05 mM dCTP-Cy3, 2 U of Taq polymerase, and 5 pM each Ngo13Fw forward and MS38R reverse primers (*see* Subheading 2.2.1). PCR conditions are as follows; first denaturation at 94 °C for 4 min and 35 cycles each consisting of second denaturation at 94 °C for 30 s, annealing at 52 °C for 30 s, extension at 72 °C for 1 min 30 s, and final extension at 72 °C for 5 min in that order.

2. Immobilize this Cy3-labeled linear DNA on a Au nanowire electrode as described in Subheading 3.1.3, **step 1**.

3.2.2 Optimization of Parameters for Electrochemical DNA Release

1. Immerse a Au nanoinjector on which Cy3-labeled linear DNA is immobilized in a 1 M aqueous KH_2PO_4 solution in which a Pt wire and a mercury sulfate electrode are immersed.

2. Find a moderate reduction potential that cleaves the Au-S bonds between the nanoinjector and the DNA. Apply electric potentials of −0.7, −0.8 and −0.9 V for 5 min to each Au nanoinjector. To remove the DNA that is detached but physically adsorbed on

Au nanowires, immerse three Au nanoinjectors in a 1 M KH_2PO_4 solution and then take them out from the solution. Repeat this three times. Measure the DNA remaining on the nanoinjectors by observing their fluorescence intensity. The least negative potential that leads to complete detachment of the DNA (usually -0.8 V) is selected as the optimal magnitude for a voltage pulse (*see* **Note 10**).

3. To find the shortest duration for a voltage pulse, the optimal magnitude of a voltage pulse is applied for 1, 2 and 3 min to each Au nanoinjector. To remove the DNA that is detached but physically adsorbed on Au nanowires, immerse three Au nanoinjectors in a 1 M KH_2PO_4 solution and then take them out from the solution. Repeat this three times. Measure the DNA remaining on the nanoinjectors by observing their fluorescence intensity. The shortest duration that leads to complete detachment of the DNA (usually 2 min) is selected to be the optimal duration for a voltage pulse (*see* **Note 10**).

3.3 DNA Delivery into the Nucleus of a Single Cell

A Au nanoinjector with immobilized DNA is inserted deep into a specific intracellular region of a cell. An optimized voltage pulse is applied to the nanoinjector to release the DNA. The nanoinjector is then withdrawn from the cell and expression of the delivered gene is investigated. As an example, DNA coding for a fluorescent protein is delivered into a cell nucleus and it is observed whether fluorescent protein is produced in the cell in order to confirm successful delivery.

3.3.1 DNA Delivery

1. Grow SK-N-SH cells on gelatin-coated glass slides (Subheading 2.3.1, **step 3**) in a culture dish.

2. Mount a Au nanoinjector with immobilized DNA on a stand fixed on a three-axis microstage. Adjust the angle of the nanoinjector to lay the entire Au nanowire almost parallel (within 5°) to the stage of an optical microscope (*see* **Note 11**).

3. Under optical monitoring, immerse the nanoinjector in the culture medium where a glass slide on which cells are attached is placed.

4. Adjust the position of the nanoinjector to a few μm above the floor of the slide.

5. Immerse Pt wire and mercury sulfate electrodes in the culture medium, as shown in Fig. 2.

6. Adjust the microscope focus on the region of the cell into which DNA will be delivered (*see* **Note 12**).

7. Control the position of the nanoinjector to approach the side of the cell and to penetrate the cell membrane as shown in Fig. 3 (*see* **Notes 13** and **14**).

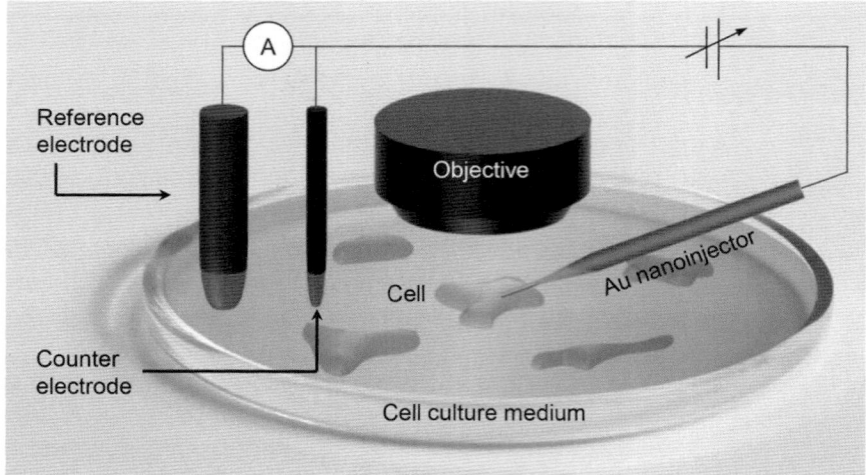

Fig. 2 Schematic illustration of the Au nanoinjector system for electrotriggered delivery of exogenous materials into a single cell (Adapted with permission from [10]. Copyright 2013 American Chemical Society)

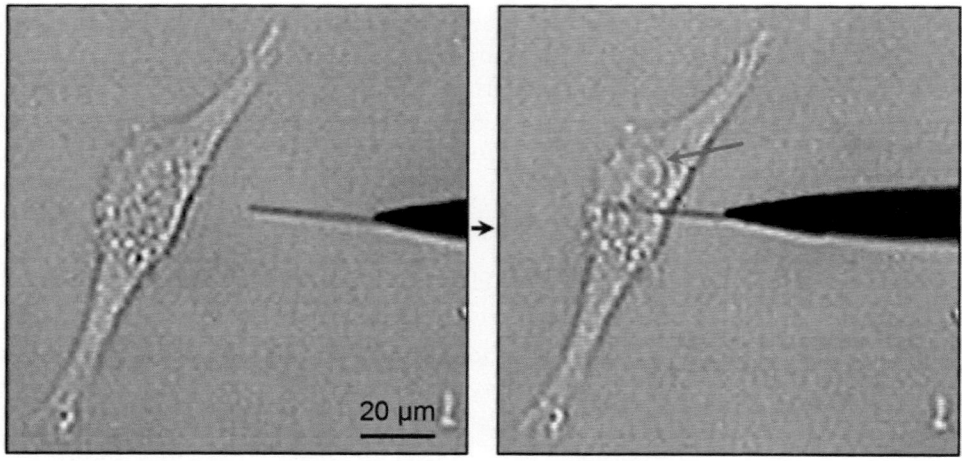

Fig. 3 Optical images showing the penetration of a Au nanoinjector into a cell nucleus. The *red arrow* denotes the nuclear membrane (Adapted with permission from [10]. Copyright 2013 American Chemical Society)

8. Apply a voltage pulse to the Au nanoinjector.

9. After the termination of the voltage pulse, withdraw the nanoinjector from the cell.

3.3.2 Test of Cell Viability

1. Add trypan blue solution into the culture medium (final concentration 0.03 % w/v).

2. Monitor the cell by microscopy for 7 h and check whether entry of trypan blue and/or changes in cell morphology occur.

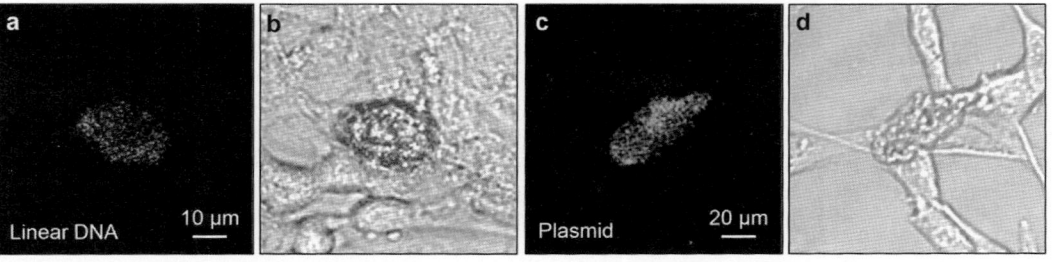

Fig. 4 Fluorescence (**a**, **c**) and optical (**b**, **d**) images of SK-N-SH cells after intranuclear delivery of linear DNA (**a**, **b**) or plasmid DNA coding for EGFP (**c**, **d**) by using Au nanoinjectors (Adapted with permission from [10]. Copyright 2013 American Chemical Society)

3.3.3 Quantification of DNA Delivered into a Cell

1. Prepare 23S ribosomal DNA of *N. gonorrhoeae* by PCR (*see* Subheading 2.2.1).

2. Deliver the DNA into cells by using Au nanoinjectors.

3. Extract DNA from all the cells on a given slide with TRIzol®-LS reagent according to the manufacturer's instructions.

4. Quantify the delivered DNA by performing qPCR on the extracted DNA with Ngo13Fw forward primer and Ngo3R reverse primer (*see* Subheading 2.2.2, **item 3**) with conditions as follows; 40 cycles of 95 °C for 20 s, 52 °C for 15 s, and 72 °C for 20 s.

5. Check the specificity of amplification by monitoring a melting curve from 72 to 95 °C following the final cycle of PCR.

6. Calculate the absolute amount of DNA delivered using a standard curve obtained from a dilution series (1.5, 7.5, and 15 ng of DNA).

3.3.4 Expression of DNA Delivered into a Cell

1. Grow cells for 48 h after delivering thiolated linear EGFP-coding DNA or EGFP-coding plasmid using Au nanoinjectors.

2. Observe the cells by fluorescence microscopy to check if the DNA is expressed, as shown in Fig. 4 (*see* **Note 15**).

4   Notes

1. The sapphire substrate with grown Au nanowires is attached on the side wall of a solid support which is then laid on the stage of an optical microscope. This arrangement makes the Au nanowires vertical to the stage.

2. For robust attachment of a nanowire to a tungsten tip, more than one-third of its length should be in contact with the tip.

3. If a Au nanowire bends away from the tungsten tip, probably due to static electricity, connect the tungsten tip to ground or to a piece of metal.

4. If the method in **Note 3** is not effective, soak the end of a tungsten tip in conducting paste prior to picking up a Au nanowire.

5. If the method in **Note 4** is not effective, expose the sapphire substrate to water vapor.

6. For the Au nanowire to take its place stably on the tungsten tip, immerse the middle of the tip of the combined Au nanowire–tungsten tip in a drop of conducting paste and move it back and forth up to the end of the tip. Expose this combined Au nanowire–tungsten tip to UV light for few minutes. This attaches the Au nanowire tightly to the tip.

7. The concentration of nail varnish solution influences the thickness of the insulating layer. It can be adjusted by diluting the solution with acetone.

8. For more robust fixation of the Au nanowire to a tungsten tip, prior to the insulation with nail varnish immerse the middle of the tip of the combined Au nanowire–tungsten tip in a drop of UV-curable polymer solution and move it back and forth up to the end of the tip. Expose this combined Au nanowire–tungsten tip to UV light for a few to tens of minutes depending on the thickness of the UV-curable polymer layer.

9. When immersing a Au nanowire electrode into a solution, it is recommended to enter the solution vertically to minimize physical stress on the nanowire.

10. The optimal magnitude and duration of a voltage pulse for reductive cleavage of Au-S covalent bonds are dependent on the structure of the thiolated molecule immobilized on a Au nanowire.

11. This is highly important when inserting the Au nanoinjector as deep as you want, since the nanoinjector is inserted into a cell parallel to the glass slide.

12. To easily track the cell into which the Au nanoinjector is inserted, a mark can be made on the underside of the glass slide before putting it into the cell culture dish. Cells placed near the mark can be easily tracked.

13. A Au nanoinjector should be inserted into the interior of the nucleus for the delivered DNA to be successfully expressed. When the nanoinjector touches the nuclear membrane and penetrates the nucleus, a small dent appears in the nuclear membrane as shown in Fig. 3. If DNA is delivered into the cytoplasm, only a very small portion (~0.3 %) reaches the nucleus due to digestion of DNA in the cytoplasm and failure of nuclear membrane penetration.

14. When the angle between a Au nanoinjector and the cell surface is close to 90°, the nanoinjector easily penetrates the cell membrane. Otherwise, it is apt to just bend and slide over the membrane without penetrating the cell.

15. The position of the cell into which EGFP-coding DNA is delivered can be easily found by making a mark on the underside of the glass slide (*see* **Note 12**).

References

1. Xie C, Lin Z, Hanson L et al (2012) Intracellular recording of action potentials by nanopillar electroporation. Nat Nanotechnol 7:185–190

2. Duan X, Gao R, Xie P et al (2012) Intracellular recordings of action potentials by an extracellular nanoscale field-effect transistor. Nat Nanotechnol 7:174–179

3. Shalek A, Robinson JT, Karp ES et al (2010) Vertical silicon nanowires as a universal platform for delivering biomolecules into living cells. Proc Natl Acad Sci U S A 107:1870–1875

4. Chen X, Kis A, Zettl A et al (2007) A cell nanoinjector based on carbon nanotubes. Proc Natl Acad Sci U S A 104:8218–8222

5. Yum K, Na S, Xiang Y et al (2009) Mechanochemical delivery and dynamic tracking of fluorescent quantum dots in the cytoplasm and nucleus of living cells. Nano Lett 9:2193–2198

6. Yum K, Wang N, Yu M-F (2010) Electrochemically controlled deconjugation and delivery of single quantum dots into the nucleus of living cells. Small 6:2109–2113

7. Han S-W, Nakamura C, Kotobuki N et al (2008) High-efficiency DNA injection into a single human mesenchymal stem cell using a nanoneedle and atomic force microscopy. Nanomedicine 4:215–225

8. Karra D, Dahm R (2010) Transfection techniques for neuronal cells. J Neurosci 30: 6171–6177

9. Gresch O, Altrogge L (2012) Transfection of difficult-to-transfect primary mammalian cells. Methods Mol Biol 801:65–74

10. Yoo SM, Kang M, Kang T et al (2013) Electrotriggered, spatioselective, quantitative gene delivery into a single cell nucleus by Au nanowire nanoinjector. Nano Lett 13: 2431–2435

11. Singhal R, Orynbayeva Z, Sundaram RVK et al (2011) Multifunctional carbon-nanotube cellular endoscopes. Nat Nanotechnol 6:57–64

12. Zheng XT, Chen P, Li CM (2012) Anticancer efficacy and subcellular site of action investigated by real-time monitoring of cellular responses to localized drug delivery in single cells. Small 8:2670–2674

13. Abouzeid AH, Torchilin VP (2013) The role of cell cycle in the efficiency and activity of cancer nanomedicines. Expert Opin Drug Deliv 10:775–786

14. Love JC, Estroff LA, Kriebel JK et al (2005) Self-assembled monolayers of thiolates on metals as a form of nanotechnology. Chem Rev 105:1103–1169

15. Creager SE, Hockett LA, Rowe GK (1992) Consequences of microscopic surface roughness for molecular self-assembly. Langmuir 8:854–861

16. Yoo Y, Seo K, Han S et al (2010) Steering epitaxial alignment of Au, Pd, and AuPd nanowire arrays by atom flux change. Nano Lett 10: 432–438

Improving Chromatin Immunoprecipitation (ChIP) by Suppression of Method-Induced DNA-Damage Signaling

Sascha Beneke

Abstract

Genomic DNA is always associated with proteins that modulate the accessibility of the genetic information. This chromatin is the essential structure in which all nuclear activity from regulation to replication, transcription, and repair takes place. This dynamic structure can be most efficiently analyzed by using the method of chromatin immunoprecipitation (ChIP), where application of cell-permeable cross-linkers to living cells induces covalent bridging between proteins and adjacent DNA in the nucleus. After fragmentation of the DNA, the complexed proteins are isolated by binding to specific antibodies. The attached DNA is isolated and can be analyzed. This method has been improved multiple times and adjusted to different experimental needs. This chapter describes a further advance based on the observation that the current standard method itself induces alterations in the chromatin.

Key words ChIP, Formaldehyde, Concentration, DNA breaks, Cross-linking, Efficiency, Damage signaling

1 Introduction

Chromatin immunoprecipitation is the most widespread method to analyze interactions of proteins with DNA, and has pushed forward the field of transcription and chromatin regulation enormously. Two substantially different versions of this method have been developed, one dealing with unchanged native chromatin (N-ChIP), and the other with chromatin in which proteins are fixed to DNA by a chemical reaction (X-ChIP). Whereas N-ChIP is generally used to study the stable interaction of DNA with tightly bound histones and their modification, X-ChIP is able to trap the more volatile binding to DNA of other proteins such as transcription factors, chromatin modifiers, and epigenetic regulators. For sample preparation in X-ChIP, a cell-permeable cross-linking molecule is applied to living cells and induces the covalent bridging of proteins or nucleic acids to their respective neighbors. In this way, nuclear proteins bound to DNA are efficiently cross-linked to their

Ronald Hancock (ed.), *The Nucleus*, Methods in Molecular Biology, vol. 1228,
DOI 10.1007/978-1-4939-1680-1_7, © Springer Science+Business Media New York 2015

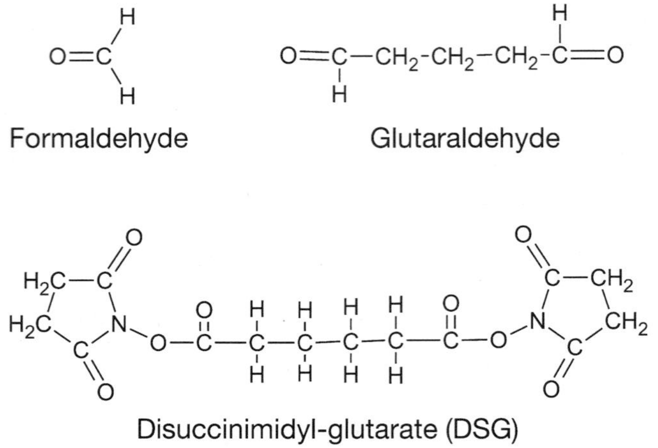

Fig. 1 Structure of three commonly used fixatives for ChIP. Formaldehyde is used most often, glutaraldehyde and disuccinimidyl-glutarate (DSG) rarely. Formaldehyde cross-links directly interacting molecules, whereas glutaraldehyde and DSG, due to the linker between the reactive groups, can bridge DNA to proteins further away, i.e., as part of a protein complex

interaction partner. Cross-linkers can be short (i.e., formaldehyde) to trap direct interactions, or display a longer stretch between reactive groups (i.e., glutaraldehyde, or N-hydroxysuccinimide-esters such as disuccinimidyl glutarate) to covalently attach more distantly located proteins within complexes (Fig. 1). Subsequently, the reaction is stopped and chemicals are washed out. Finally the cells are lysed, the chromatin is fragmented, and protein concentration is adjusted in order to achieve comparable and specific immuno-precipitation (Fig. 2). Usually, formaldehyde at a concentration of 1 % is applied to induce mild cross-linking.

However, due to their reactive nature all cross-linkers generate a variety of adducts on proteins and nucleic acids which theoretically could elicit a DNA-damage response. One immediate early damage signal is poly(ADP-ribosyl)ation (PARylation), a chromatin-restructuring polymer which is synthesized transiently after formation of DNA breaks or kinks, mainly by poly(ADP-ribose)polymerase-1 (PARP1) [1, 2]. Another marker for DNA breaks is the phosphorylation of the histone variant H2AX (termed gamma-H2AX) by the PI3K-related initiator-kinases ATM, ATR, and DNA-PK [3, 4]. Strong gamma-H2AX foci mark persistent double-strand breaks and can be detected even hours after geno-toxic impact. Indeed, both signals are detectable in ChIP-fixed cells, suggesting that the low concentration of formaldehyde uti-lized in the standard ChIP method is unable to abrogate the enzy-matic activity of DNA damage sensors. PARylation is induced weakly in the fixation step of X-ChIP, but increases massively in the subse-quent washes [5]. Application of PARP inhibitors throughout the

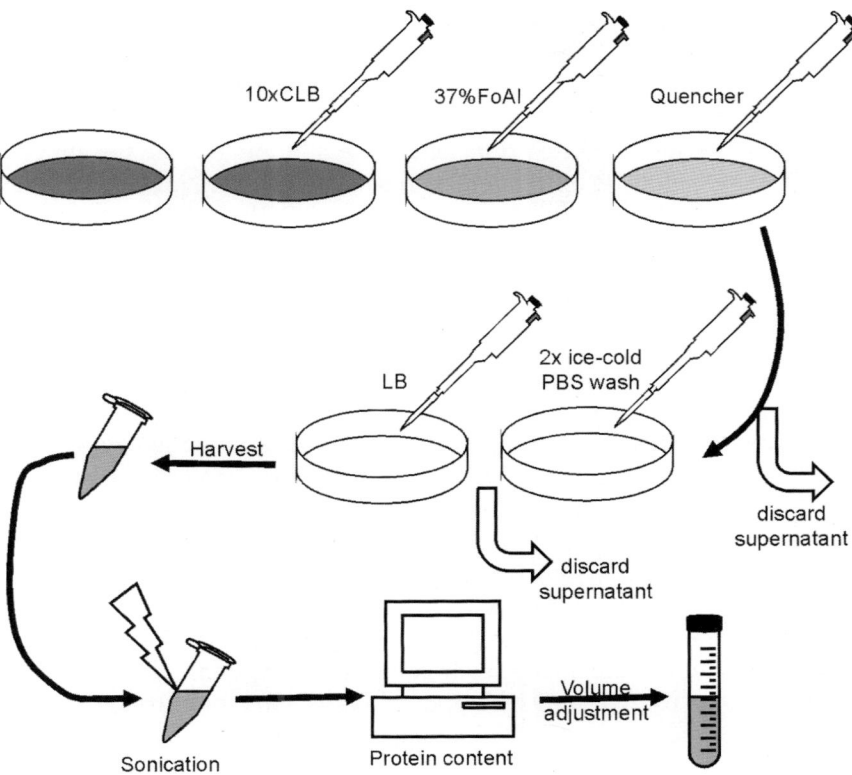

Fig. 2 Flowchart for fixation and sonication. 10× CLB (cross-linking buffer) is added to the cells followed by 1/9 volume of 37 % formaldehyde for 10 min at room temperature, and then quencher to stop the cross-linking process. The supernatant is replaced by ice-cold PBS and the fixed cells are washed twice on ice. Cells are lysed on the plate and scraped off. After sonication, the protein content is adjusted to 1 μg/μL with LB and the chromatin solution can be stored at 4 °C

entire process can suppress PARylation, but it does not impact on gamma-H2AX formation. Blocking of the phosphorylation reaction is only achieved by increasing the formaldehyde concentration to 3.7 %, probably due to the complete inhibition of all enzymatic activity. Both these treatments change ChIP efficiency, but this depends on the interacting partners, i.e., the protein and its binding site in the genome [5].

Reasoning from these observations, we have developed an alternative protocol based on an established X-ChIP method [6]. We add to the cell growth medium a HEPES cross-linking buffer which keeps the pH in a suitable range. Afterwards, formaldehyde is added to a concentration of 3.7 %, abrogating all enzymatic activity. Subsequently, the fixed cells are washed and lysed on ice. In this way, damage-signaling pathways are suppressed despite the presence of DNA breaks. This high concentration of formaldehyde has the disadvantage that it induces massive cross-fixation which hinders fragmentation of DNA by shearing chromatin, which is a necessary step in order to achieve specific precipitation of protein-DNA complexes. If fragmentation is too great, the subsequently isolated DNA is too small to yield a proper signal in the PCR detection step, while if fragmentation is too low precipitated high molecular weight DNA could contain sequences detectable in the later PCR step despite no specific protein binding at the site aimed for. Thus, if 3.7 % formaldehyde is used for fixation, proper antibody controls and careful determination of PCR conditions are mandatory. If these precautions are followed, this method preserves the natural composition of chromatin better than the standard version, since method-induced signaling events are suppressed. Regardless of the detection method that follows the fixation step, i.e., basic ChIP with subsequent PCR, high-throughput ChIP-on-chip, ChIP-Seq, or other, damage signaling has to be suppressed as this changes chromatin structure, either masking proper interaction of proteins with DNA by reducing their presence at the respective site, or by yielding false-positives by recruitment of proteins (own unpublished results).

This chapter describes our improved protocol for X-ChIP followed by PCR detection without the use of sophisticated equipment, except for the chromatin fragmentation step. Before starting a large-scale setup, a pre-experiment should be performed to determine the optimal conditions for chromatin fragmentation as well as for PCR amplification, and this is included in the protocol described here. As stated above, chromatin fragmentation efficiency can vary between different cell lines, so optimal settings have to be determined for each one. This can be achieved by taking samples during the course of shearing and analyzing their DNA on an agarose gel after purification. There are currently two major ways to fragment chromatin: one is by sonication, and the other by using restriction enzymes. We use the Bioruptor sonicator (Diagenode) in which samples are fragmented in a cooled water-bath by a sequence of cycles with fixed sonication intensity, and therefore working in a cold-room and on ice is no longer needed. An alternative protocol is described in **Note 1**. Sonication of samples cross-linked with the standard fixation in 1 % formaldehyde yields visible fragments below 1–2 kb in length in optimum conditions, with the highest concentration at around 500 bp (Fig. 3, lanes 1–3). Using 3.7 % formaldehyde will never produce such efficient fragmentation and there is always a portion of large chromatin fragments left, giving a DNA smear in the gel from the well down to the bottom (Fig. 3, lanes 4–6).

To analyze ChIP efficiency we use semiquantitative PCR. Another frequently applied method is Real-Time Q-PCR, but this needs special equipment and even may not be as accurate as normal PCR, since Q-PCR with SYBR green to detect double stranded DNA also includes any background smear in the measurement. Thus, agarose gel electrophoresis of amplification reactions after Q-PCR should be mandatory, although it is not always done. Semiquantitative PCR has the disadvantage that

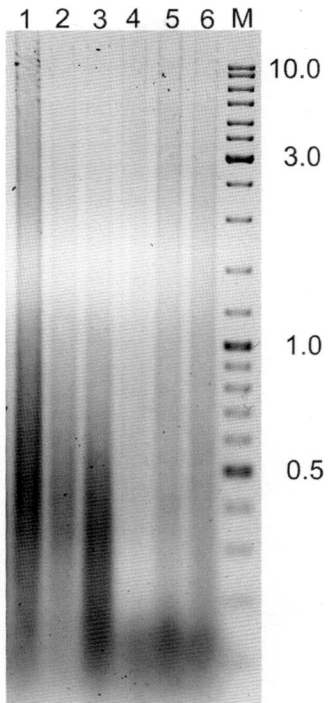

Fig. 3 Fragmentation efficiency is reduced after fixation with 3.7 % formaldehyde. The agarose gel (0.8 %) shows DNA from chromatin samples fixed with 1 % (*lanes 1–3*) or 3.7 % (*lanes 4–6*) formaldehyde. Each set of three lanes represent 10, 20, or 30 sonication cycles for each fixation condition, respectively. *Lane M* contains a base-pair marker with the indicated sizes. Whereas DNA from samples fixed in 1 % formaldehyde displays the expected size decrease with increased cycle numbers with an optimum at 20 cycles, the size of DNA from samples fixed in 3.7 % formaldehyde is rather stable and shows only little fragmentation. The best results are obtained after 50 cycles (data not shown)

conditions have to be established in which the PCR is not only specific, but also gives a linear response. Analysis of PCR products by agarose gel electrophoresis has the advantage that you actually see the specificity of the amplification reaction and in addition, only the intensity of a specific fragment can be measured without the nonspecific background. We use the very efficient KOD-polymerase for PCR, as it displays superb amplification rates. Selection of oligonucleotide sequences for primers is another crucial parameter in PCR. Primers should be at least 18 nucleotides long for specificity reasons, and an annealing temperature T_A above 52 °C is desirable. In Q-PCR a 100 bp amplicon is standard, which is the lower limit in semiquantitative PCR due to the resolution capacity of agarose gel electrophoresis. Here, amplicon lengths of 120 bp up to 500 bp can be separated well and are still of a "ChIP-able" size. To determine the optimal

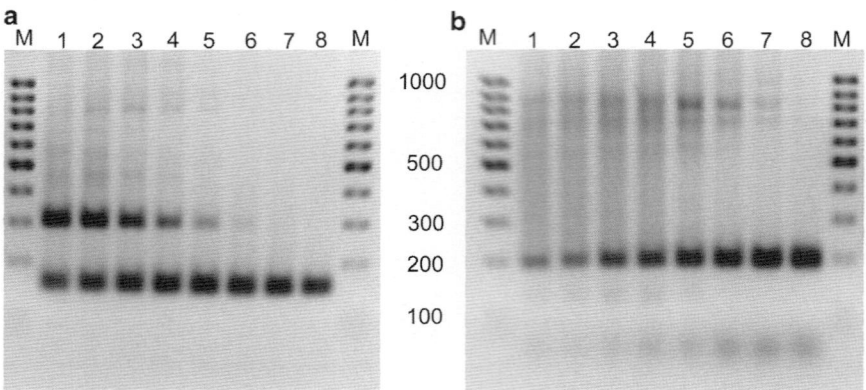

Fig. 4 Optimization of PCR conditions. Using a gradient cycler, the annealing temperature was increased by 1.2 °C steps between different samples of the promoter sequence tested (panel (**a**) = gene 1; panel (**b**) = gene 2), ranging in panel (**a**) from 54.8 °C in *lane 1* to 63.2 °C in *lane 8*, and in panel (**b**) from 52.6 °C in *lane 1* to 61 °C in *lane 8*. The specificity increases with increasing annealing temperature, reaching a maximum at 62 °C in panel (**a**) and 61 °C in panel (**b**). Of note, the amount of amplification product decreases again at the highest temperature in panel (**a**), probably due to less efficient primer binding. *Lane M* contains a base-pair marker with the indicated sizes

conditions for semiquantitative PCR, the best starting material is the purified input DNA. Three major parameters have to be optimized for ChIP-PCR. First, the best annealing temperature has to be determined. Starting from the calculated T_M, annealing temperature should be increased and decreased in 1 °C steps over 10 °C range (Fig. 4). As a rule of thumb, the higher the annealing temperature, the more specific is the amplification. A gradient cycler is convenient for this, but of course it can also be done in different single PCRs. Second, the amount of background products can be influenced by the concentration of Mg^{2+} ions in the buffer, which can be adjusted as well if necessary. Finally, optimal cycling times should be determined to reach linearity of amplification in the tested conditions to avoid saturation. The examples in Figs. 4, 5, and 6 show the differences in sample processing and subsequent results between cells fixed with 1 or 3.7 % formaldehyde.

2   Materials

1. Cell growth medium: Dulbecco's Modified Eagle's Medium (DMEM) with high glucose, 100 μg/mL streptomycin, 100 U/mL penicillin, and 10 % fetal calf serum (FCS) for HeLa S3 cells.

2. 10× cross-linking buffer (10× CLB): 50 mM HEPES, pH 8.0, 100 mM NaCl, 0.5 mM EGTA pH 8.0.

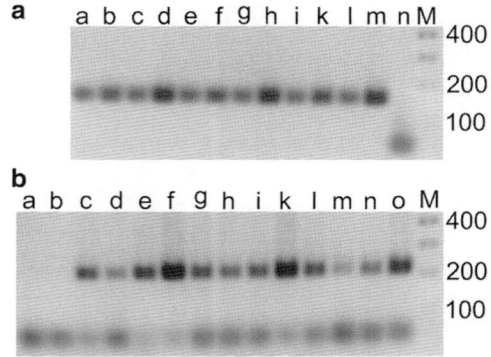

Fig. 5 ChIP PCR amplification of two different promoter sequences. Panel (**a**) displays the amount of PCR product from three independent ChIP experiments, including independent chromatin preparations, testing for a transcription factor binding site in gene 1. Compared were ChIP efficiencies of 3.7 and 1 % formaldehyde fixations side-by-side. The loading scheme was as follows: three consecutive settings with input-PCR 3.7 % (*lanes a,e,i*), ChIP-PCR 3.7 % (*lanes b,f,k*), ChIP-PCR 1 % (*lanes c,g,l*), input-PCR 1 % (*lanes d,h,m*) with a ChIP-PCR antibody control in *lane n*. Lane M contains a base-pair marker with the indicated sizes. Panel (**b**) displays the amount of PCR product from three independent ChIP experiments, including independent chromatin preparations, testing for presence of the same transcription factor as in (**a**) to a binding site in gene 2. Compared were ChIP efficiencies of 3.7 and 1 % formaldehyde fixations side-by-side. Loaded in *lanes a* and *b* were ChIP-PCR from antibody controls of 3.7 and 1 % chromatin fixation, respectively, followed by three consecutive settings with input-PCR 3.7 % (*lanes c,g,l*), ChIP-PCR 3.7 % (*lanes d,h,m*), ChIP-PCR 1 % (*lanes e,i,n*), input-PCR 1 % (*lanes f,k,o*). Lane M contains a base-pair marker with the indicated sizes. The amount of amplified fragment is comparable between the independent experiments

3. Formaldehyde: we do not have a preferred supplier. A fresh bottle is best since formaldehyde slowly polymerizes which reduces its effective concentration, but a bottle can still be used one year after opening.

4. Quenching solution: 1.25 M glycine in 1× CLB.

5. Lysis buffer (LB): 50 mM Tris–HCl pH 8.0, 10 mM EDTA pH 8.0, 1 % SDS, 1× protease inhibitor cocktail, 1× phosphatase inhibitor cocktail (optional) (*see* **Note 2**).

6. Cell scraper.

7. Dilution buffer (DB): 150 mM NaCl, 20 mM Tris–HCl pH 8.0, 2 mM EDTA pH 8.0, 1 % Triton X-100, 1× protease inhibitor cocktail, 1× phosphatase inhibitor cocktail (optional) (*see* **Note 2**).

8. Mix buffer (MB): 9 parts DB + 1 part LB.

9. Protein G-agarose beads: 50 % (v/v) slurry, capacity 10–15 mg human IgG/mL (Sigma or other).

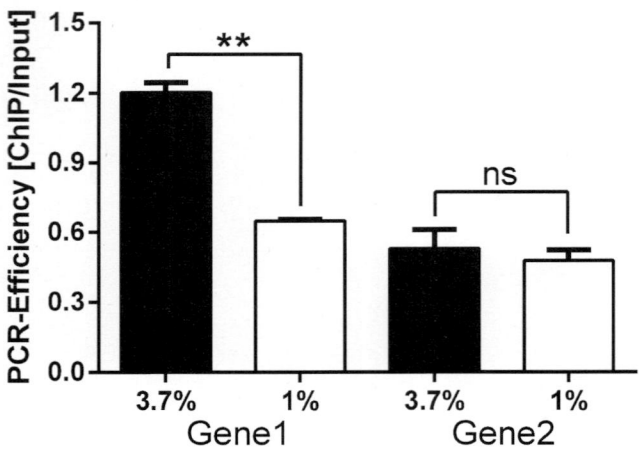

Fig. 6 Evaluation of signal intensities from Fig. 5. The signal of each specific PCR product was measured with ImageJ and normalized to that of the antibody control. Subsequently, the ratios of the respective ChIP/input signals were calculated and displayed as bar diagrams using the GraphPad Prism5 program. ChIP-efficiency is substantially different between 3.7 and 1 % formaldehyde fixations for the transcription factor binding to the Gene1 promoter, whereas there is no difference for the transcription factor binding to the Gene2 promoter. Thus, fixation conditions impact on ChIP efficiency. Of note, amplification of input DNA was not significantly different between the two fixation conditions with regard to yield

10. Wash buffer (WB): 150 mM NaCl, 20 mM Tris–HCl pH 8.0, 2 mM EDTA pH 8.0, 1 % Triton X-100, 0.1 % SDS, 1× protease inhibitor cocktail, 1× phosphatase inhibitor cocktail (optional) (*see* **Note 2**).

11. Final wash (FW): 500 mM NaCl, 20 mM Tris–HCl, pH 8.0, 2 mM EDTA, pH 8.0, 1 % Triton X-100, 0.1 % SDS, 1× protease inhibitor cocktail, 1× phosphatase inhibitor cocktail (optional) (*see* **Note 2**).

12. Elution buffer (EB): 100 mM $NaHCO_3$, 1 % SDS.

13. Solutions for ChIP: 5 mg/mL sheared salmon sperm DNA (SSSD), 10 mg/mL BSA.

14. Solutions for cross-link reversal and digestion: 20 mg/mL ribonuclease A (RNase A), 20 mg/mL proteinase K (PK), 5 M NaCl.

15. Solutions for DNA isolation: Tris-buffered phenol, chloroform–isoamyl alcohol 29:1 or 49:1. (CIA), 3 M Na acetate, pH 5.5, 100 % ethanol, 70 % ethanol, nuclease-free water.

16. KOD-Hot Start PCR kit.

17. PCR primers.

18. ThermoShaker (variable speed and variable temperature microtube mixer).

19. Sonicator: we use the Bioruptor (Diagenode, Belgium). For an alternative *see* **Note 1**.

20. Rocking platform.

21. Refrigerated microcentrifuge.

22. Solutions for your preferred assay of protein concentration (Bradford, Lowry, BCA, etc.) or simply use a spectrophotometer.

23. Material for agarose gel electrophoresis.

3 Methods

Pre-experiment

Start an initial small-scale pre-experiment as described below for each cell line to be used for ChIP. You have to do this because (1) chromatin fragmentation efficiency may vary between different cell lines so you have to find the optimal setting to yield usable chromatin for ChIP, and (2) you have to define the optimal conditions for the subsequent PCR amplification. Usually, one 10 cm dish with semiconfluent cells yields enough material for pre-experiments and for about three individual ChIPs. In the following Subheading 3.1, an experiment will be described using cells from one 15 cm dish which yields enough material for input and ChIPping with at least three different antibodies.

3.1 Growing Cells

1. Seed the cells at a density equivalent to 30–40 % confluence, which for the HeLa S3 cell strain is about $1–2 \times 10^7$ cells in a 15 cm dish or $5–6 \times 10^6$ cells in a 10 cm dish, at least 24 h ahead of starting experiments in order to achieve complete adaptation to the environment and to achieve about 50 % confluence after 24 h. Use 18 mL of the growth medium of your choice.

2. Grow for at least 24 h in the incubator.

Treat the cells according to the experimental design. Always compare control with treated cells side-by-side.

3.2 Cross-Linking

1. The protocol for X-ChIP described below is based on [6] but uses a concentration of 3.7 % formaldehyde instead of 1 % for cross-linking.

2. Start the cross-linking protocol by adding 1/9 volume of 10x CLB to the culture medium, i.e., 2 mL to 18 mL medium. Swirl the plate until the CLB and medium are completely mixed.

3. Add 1/9 volume (2.3 mL) of 37 % formaldehyde to the mixture of 10× CLB + medium, yielding a 3.7 % final concentration of formaldehyde (*see* **Note 3**).

4. Swirl the plate until completely mixed (the medium will turn orange).

5. Incubate for 10 min at room temperature on a swirling or rocking plate (speed should be low).

6. Add 2.0 mL of quenching solution to the fixed cells (the medium will turn yellow) to reach 0.1 M glycine. This step quenches (inactivates) any residual formaldehyde. Swirl the plate until the solution is completely mixed.

7. Incubate for 2 min on a swirling or rocking platform (speed should be low).

8. Replace the supernatant by 10 mL of ice-cold PBS and incubate for 3 min on ice on a swirling or rocking platform (speed should be low) (*see* **Note 4**).

9. Repeat this step once.

10. Discard the supernatant and replace by 1 mL of cooled LB (*see* **Note 5**). Be sure to distribute the LB evenly on the dish.

11. Lyse the cells for 3–5 min at 4–8 °C in the refrigerator.

12. Tilt the dish slightly, scrape off the lysed cells, and harvest them by pipetting. This will yield about 1.4 mL of lysate.

3.3 Chromatin Fragmentation and Volume Adjustment

As stated above, the efficiency of chromatin fragmentation can vary between different cell lines, so optimal settings have to be determined for each one. We use the Bioruptor with the following settings for HeLa S3 cells: 30 cycles with 30 s sonication/30 s off time and high intensity (for alternatives, *see* **Note 1**). To monitor shearing take a sample every ten cycles, isolate DNA, and analyze by agarose gel electrophoresis (Fig. 3). After sonication, concentration of the samples has to be determined and adjusted.

1. Cool the sonicator's water bath to 4 °C.

2. Load the samples in the tubes and with the volumes recommended by the manufacturer. If necessary, split the sample into several tubes.

3. Sonicate the samples with the settings determined previously to produce an appropriate fragmentation pattern.

4. Determine the absorbance at 260 nm relative to that of LB alone. Alternatively, absorbance at 280 nm can be measured relative to LB (*see* **Note 6**).

5. Bring the lysates to an identical absorbance by adding LB.

6. This solution can be stored up to several months at 4 °C (*see* **Note 7**).

3.4 Antibody Binding

We suggest to include a proper antibody control for each ChIP, for example an antibody with similar features (raised in the same species, of monoclonal or polyclonal origin), but which binds to a target located in a different cellular compartment such as the cytoplasmic membrane.

1. Bring the lysate to room temperature.

2. Spin down 220 µL of lysate for 5 min at maximum speed in a microcentrifuge.

3. Take 200 µL of supernatant, transfer it to a 2 mL tube, and add 1,800 µL of DB.

4. Add 2 µg of antibody and rotate for at least 6 h (overnight is better) at 4 °C.

3.5 Protein G Preparation

1. Resuspend the protein G bead slurry thoroughly (*see* **Note 8**).

2. Add 50 µL of the slurry of protein G beads to 500 µL of MB.

3. Add BSA to 100 µg/mL and SSSD to 500 µg/mL (*see* **Note 9**).

4. Rotate overnight at 4 °C.

5. Spin down the beads at $1,000 \times g$ for 5 min at 4 °C in a microcentrifuge.

6. Discard the supernatant.

7. Resuspend the beads in 500 µL of MB.

8. Take 50 µL of the slurry of beads for each IP (*see* **Note 10**).

3.6 Chromatin Immunoprecipitation

1. Add 50 µL of the suspension of protein G beads in MB to each IP.

2. Rotate for 2 h at 4 °C.

3. Spin down the beads at $1,000 \times g$ for 10 min at 4 °C in a microcentrifuge.

4. Discard the supernatant carefully.

5. Resuspend the beads in 500 µL of WB.

6. Repeat the last three steps 2× (i.e., three3 washes total).

7. Spin down the beads at $1,000 \times g$ for 5 min at 4 °C in a microcentrifuge.

8. Resuspend the pellet in 500 µL of FW.

9. Spin down the beads at $1,000 \times g$ for 5 min at 4 °C in a microcentrifuge.

10. Discard the supernatant by decantation.

11. Spin down the beads at $1,000 \times g$ for 5 min at 4 °C in a microcentrifuge.

12. Very carefully remove the residual supernatant by pipetting and discard it.

3.7 Reversal of Cross-Links and Elution

1. Resuspend the beads in 250 µL of EB.

2. Place the tube in a ThermoShaker with 600 rpm setting and 37 °C.

3. Add 6.3 µL (2.5 % sample volume) of RNaseA to reach a concentration of 500 µg RNaseA/mL.

4. Incubate for 30 min at 37 °C with shaking to degrade RNA.

5. Add 6.3 μL (2.5 % sample volume) of PK solution to reach a concentration of 500 μg PK/mL.

6. Incubate for 1 h at 56 °C to degrade proteins.

7. Add 10.5 μL (4 % sample volume) of NaCl solution to reach a final concentration of 200 mM NaCl.

8. Increase the temperature to 65 °C and incubate for at least 4 h (overnight is preferable) for reversion of cross-links and further protein degradation.

9. Spin down for 5 min at full speed in a microcentrifuge.

10. Transfer the supernatant to a 1.5 mL microtube.

Purify the DNA by standard phenol–chloroform extraction.

3.8 Preparation of Input DNA

This should done side-by-side with the ChIP samples in order to get comparable yields.

1. Bring 100 μL of lysate to room temperature.

2. Place in a ThermoShaker with 600 rpm setting and 37 °C.

3. Add 2.5 μL (2.5 % sample volume) of RNaseA to reach a concentration of 500 μg/mLRNase.

4. Incubate for 30 min at 37 °C with shaking to degrade RNA.

5. Add 2.5 μL (2.5 % sample volume) of PK to reach a concentration of 500 μg/mL PK.

6. Incubate for 1 h at 56 °C to degrade proteins.

7. Add 4.2 μL (4 % sample volume) of NaCl to reach a final concentration of 200 mM NaCl.

8. Increase the temperature to 65 °C and incubate for at least 4 h (overnight is preferable) for reversion of cross-links and further protein degradation.

9. Spin down 5 min at full speed in a microcentrifuge.

10. Transfer the supernatant to a 1.5 mL microtube.

11. Purify the DNA by standard phenol–chloroform extraction.

3.9 DNA Extraction with Phenol–Chloroform

1. Add 100 μL of phenol to the sample (0.5–1× the sample volume).

2. Vortex for 10 s.

3. Spin down for 5 min at full speed in a microcentrifuge.

4. Take out carefully most of the upper phase containing the DNA and transfer this to a new microtube. Be sure NOT to transfer any liquid from the interphase (denatured proteins etc.) or lower phase (phenol).

5. Take out the phenol phase carefully and discard it.

6. Add 40 µL of nuclease-free water to the interphase material remaining in the tube.

7. Vortex for 10 s.

8. Spin down for 5 min at full speed in a microcentrifuge.

9. Take out carefully most of the upper phase containing DNA. Be sure NOT to transfer any liquid from the lower phase (phenol).

10. Transfer to the tube containing the first DNA supernatant.

11. Add to this tube 100 µL of CIA.

12. Vortex for 10 s.

13. Spin down for 5 min at full speed in a microcentrifuge.

14. Take out carefully most of the upper phase containing the DNA and transfer to a new tube. Be sure NOT to transfer any liquid from the lower phase (CIA).

15. Determine the total volume of the DNA-containing solution.

16. Add 1/10th of this volume of 3 M Na acetate, pH 5.2 to the sample and mix by inverting.

17. Add three sample volumes of 100 % ethanol at −20 °C to the sample and mix by inverting.

18. Precipitate the DNA by incubating for 2 h at −20 °C (*see* **Note 11**).

19. Spin down the DNA by centrifugation for 1 h at 4 °C and full speed in a microcentrifuge.

20. Discard the supernatant carefully by pipetting.

21. Wash the DNA pellet 1× with 150 µL of 70 % ethanol at −20 °C.

22. Spin down the DNA by centrifugation for 30 min at 4 °C and full speed in a microcentrifuge.

23. Discard the supernatant carefully by pipetting.

24. Dry the DNA by vacuum in a SpeedVac. Be sure not to over-dry, usually 2–3 min are enough (*see* **Note 12**).

25. Resuspend the DNA in nuclease-free water (100 µL for the input sample, 25 µL for the ChIP sample).

26. Store the samples at −20 °C.

3.10 Polymerase Chain Reaction and Evaluation

We use semiquantitative PCR with KOD HotStart polymerase to analyze ChIP efficiency. After optimal conditions have been set (T_A, Mg^{2+}, cycling times), ChIP and input DNA samples can be amplified. A 10 µL-volume PCR reaction containing 0.5 µL of input-DNA or at least 1 µL of ChIP-DNA, respectively, yields enough material for analysis. In order to generate comparable results between experiments, input- and ChIP-PCRs have to be performed in parallel. After PCR, separate the amplified fragments of input-PCR and ChIP-PCR on a 2.5 % agarose gel with the treated and control samples side-by-side. In this way, the intensities of the products of ChIP-PCRs can be evaluated and compared directly with those of the respective input-PCRs. Two examples of ChIP/input PCR-products generated from independent triplicate experiments are shown in Fig. 5. In order to analyze the PCR efficiency, images should be stored as TIF files with at least 600 dpi if possible. For evaluation of band intensities as shown in Fig. 6, use either a commercially available program or the ImageJ freeware.

4   Notes

1. Alternatively, sonication can be performed using a tip-sonicator with the samples in an ice-bath. We have used the Bandelin Sonoplus HD-070 with the MS73 tip (Bandelin, Berlin, Germany). Cell suspensions are sonicated on ice for 10 s at 50 % output with a 2-min pause; the optimal number of these cycles has to be determined for each cell line. This method has the advantage that larger sample volumes can be processed, but shearing may not always be completely comparable to that produced by the Bioruptor.

2. Protease and phosphatase inhibitor cocktails should be added to solutions directly before use.

3. Formaldehyde is volatile and thus may be best used in a ventilated hood. Adhere to regulations about formaldehyde usage and disposal in your country.

4. Pretreatment of cells with cytotoxic substances (e.g., DNA-damaging agents or cell cycle or metabolic inhibitors) can lead to detachment of cells from the culture dish. In this case, harvest any floating cells in the successive supernatants (CLB, formaldehyde, quencher, PBS) and pool them with the cells scraped from the dish before lysis.

5. LB contains 1 % SDS, which precipitates on ice. Prepare and cool LB solution just before use.

6. LB solution including inhibitors hampers measurement of protein concentration in some cases. Therefore we recommend measuring the relative absorption at 260 nm, which is correlated with the concentration of chromatin. As an alternative, relative absorption at 280 nm can be measured and compared to a standard curve with increasing concentrations of BSA in LB.

7. In our hands, ChIP works well even with samples stored for 18 months.

8. Make sure that everything is suspended, which may take some time.

9. Do not add BSA if you use antibodies raised against antigens coupled to BSA! This is important, as otherwise the antibody may bind to the BSA and not to the antigen.

10. Residual beads can be stored for at least up to 1 week.

11. This step can alternatively be done for 1 h at −80 °C, but this can induce salt precipitation and additionally, samples may freeze.

12. Alternatively, samples can be dried by short incubation at 45–50 °C. Usually 3–5 min is enough.

References

1. Kraus WL, Hottiger MO (2013) PARP-1 and gene regulation: progress and puzzles. Mol Aspects Med 34:1109–1123

2. Beneke S (2012) Regulation of chromatin structure by poly(ADP-ribosyl)ation. Front Genet 3:161–116

3. Firsanov DV, Solovjeva LV, Svetlova MP (2011) H2AX phosphorylation at the sites of DNA double-strand breaks in cultivated mammalian cells and tissues. Clin Epigenet 2: 283–297

4. Scully R, Xie A (2013) Double strand break repair functions of histone H2AX. Mutat Res 750:5–14

5. Beneke S, Meyer K, Holtz A et al (2012) Chromatin composition is changed by poly(ADP-ribosyl)ation during chromatin immunoprecipitation. PLoS One 7:e32914

6. Martens JH, O'Sullivan RJ, Braunschweig U et al (2005) The profile of repeat-associated histone lysine methylation states in the mouse epigenome. EMBO J 24:800–812

Chapter 8

Purification of Specific Chromatin Loci for Proteomic Analysis

Stephanie D. Byrum, Sean D. Taverna, and Alan J. Tackett

Abstract

Purification of small, native chromatin regions for proteomic identification of specifically bound proteins and histone posttranslational modifications is a powerful approach for studying mechanisms of chromosome metabolism. Here we detail a Chromatin Affinity Purification with Mass Spectrometry (ChAP-MS) approach for affinity purification of 1 kb regions of chromatin for targeted proteomic analysis. This approach utilizes quantitative, high resolution mass spectrometry to categorize proteins and histone posttranslational modifications co-enriched with the given chromatin region as either "specific" to the targeted chromatin or "nonspecific" contamination. In this way, the ChAP-MS approach can help define and redefine mechanisms of chromatin-templated activities.

Key words Chromatin, Purification, Proteomics, Epigenetics, Histone, Posttranslational modification

1 Introduction

Chromatin-associated proteins and epigenetic factors play key roles in the regulation of numerous chromosomal activities, such as gene transcription. The precise mechanisms by which these proteins function in the context of chromatin are largely unknown. One of the simplest questions one can ask is at what chromosomal location does a particular protein or histone posttranslational modification (PTM) exist. Most current methodologies address protein or histone PTM localization along chromosomes by chromatin immunoprecipitation (ChIP). ChIP involves using antibodies that recognize a specific protein or histone PTM to pull down the chromatin and then identifying the associated DNA [1–7]. ChIP has several limitations, particularly that the researcher must know the protein/PTM of interest to target and also have a specific antibody which recognizes it. Therefore, ChIP is by nature a biased approach that requires target information. However, a biochemical approach that would provide unbiased identification of proteins and histone

Ronald Hancock (ed.), *The Nucleus*, Methods in Molecular Biology, vol. 1228,
DOI 10.1007/978-1-4939-1680-1_8, © Springer Science+Business Media New York 2015

PTMs at a given genomic locus could provide unprecedented insight into molecular mechanisms regulating chromosomal activities.

Here, we detail a method called *Chromatin Affinity Purification with Mass Spectrometry* (ChAP-MS) to help chromatin analysis proceed in an unbiased manner [8]. ChAP-MS provides for the site-specific enrichment of a targeted ~1 kb region of a chromosome followed by unambiguous identification of associated proteins and histone PTMs using high resolution mass spectrometry. To validate this approach we isolated a native genomic locus, a region 5′ to the *GAL1* gene, from *Saccharomyces cerevisiae* and identified specifically bound proteins and PTMs in transcriptionally active and repressive conditions [8]. The method uses homologous recombination to engineer a LexA DNA binding site at the desired chromosomal location and an ectopically expressed, Protein A-tagged LexA fusion protein as the affinity handle for purification. Here we detail the methodology used to perform the ChAP-MS approach [8].

2   Materials

2.1   Cell Culture and Growth

1. Yeast growth medium: dissolve 6.7 g of yeast nitrogen base without amino acids (Sigma Y0626) and 2 g of yeast synthetic drop-out mix minus lysine and tryptophan (US Biological D9537-12) in 900 mL of distilled H_2O. Sterilize by autoclaving and cool to room temperature.

2. Add 1 mL of sterile filtered solution (80 mg/mL in distilled H_2O) of normal lysine for isotopically light media, or of $^{13}C6^{15}N2$-lysine (Cambridge Isotope Laboratories CNLM-291-H) for isotopically heavy media. Add 1 mL of ampicillin (100 mg/mL in 50 % ethanol) and 100 mL of either autoclaved 20 % (w/v) glucose or autoclaved 30 % (w/v) galactose.

3. Refrigerated centrifuge with 1 L bottles.

2.2   Reagents

1. Formaldehyde solution (37 % w/v).

2. *N*,*N*-dimethylformamide.

3. Protease inhibitor cocktail for fungal and yeast extracts (Sigma).

4. M270-epoxy Dynabeads (Life Technologies): suspend in *N,N-dimethylformamide* at 30 mg/mL, store at 4 °C.

5. Purified rabbit IgG (MP Biomedicals 55944): resuspend in distilled H_2O at 17 mg/mL, store in 100 μL aliquots at −80 °C.

6. Polyvinylpyrrolidone.

7. GelCode Blue Stain for SDS-PAGE gels (Pierce).

2.3 Solutions	1. 2.5 M glycine, autoclaved.

1. 2.5 M glycine, autoclaved.

2. 1.2 % (w/v) polyvinylpyrrolidone, 20 mM Hepes, pH 7.5.

3. 1 M ammonium sulfate, 60 mM sodium phosphate, pH 7.4.

4. Sodium phosphate: 0.1 M, pH 7.4.

5. Ammonium sulfate: 3 M solution in distilled H_2O.

6. Solutions for washing IgG-coated Dynabeads:

   100 mM glycine, pH 2.5.

   10 mM Tris, pH 8.8.

   Phosphate-buffered saline (PBS): 210.0 mg/L KH_2PO4, 9 g/L NaCl, 726.0 mg/L $Na_2HPO_4 \cdot 7H_2O$.

   PBS + 0.5 % (v/v) Triton X-100 (PBS-T).

7. Affinity purification buffer: 20 mM Hepes, pH 7.5, 1 M NaCl, 1 M urea, 2 mM $MgCl_2$, 0.1 % (v/v/) Tween-20, with and without 1/100 fungal protease inhibitor cocktail. Prepare immediately prior to use and keep at 4 °C.

8. Wash buffer: 20 mM Hepes pH 7.5, 2 mM $MgCl_2$, 10 mM NaCl, 0.1 % Tween-20.

9. 0.5 N ammonium hydroxide/0.5 mM EDTA solution: make immediately prior to use.

10. SDS-PAGE loading buffer (2×): 125 mM Tris–HCl, pH 6.8, 20 % v/v/ glycerol, 4.1 % w/v SDS, 4 % 2-v/v/ mercapto-ethanol, 0.05 % bromophenol blue. Store at −20 °C in 0.5 mL aliquots.

2.4 Equipment

1. Tissuemiser homogenizer (Fisher) or any available probe tissue homogenizer that can operate at 30,000 rpm.

2. Ball Mill (Retsch MM301) with stainless steel cylinders, Retsch 25 mL screw-top grinding jars, and Retsch 20 mm stainless steel ball bearings.

3. Sonicator (Bioruptor UCD-200, Diagenode).

4. Rocking platform.

5. Concentrator (SpeedVac).

6. Equipment for SDS-PAGE.

7. Liquid nitrogen.

3 Methods

The methodological workflow for a ChAP-MS analysis is shown in Fig. 1. The key components of this methodology are (1) a strain of *Saccharomyces cerevisiae* with an engineered LexA DNA binding site and an affinity-tagged LexA-Protein A fusion protein expressed

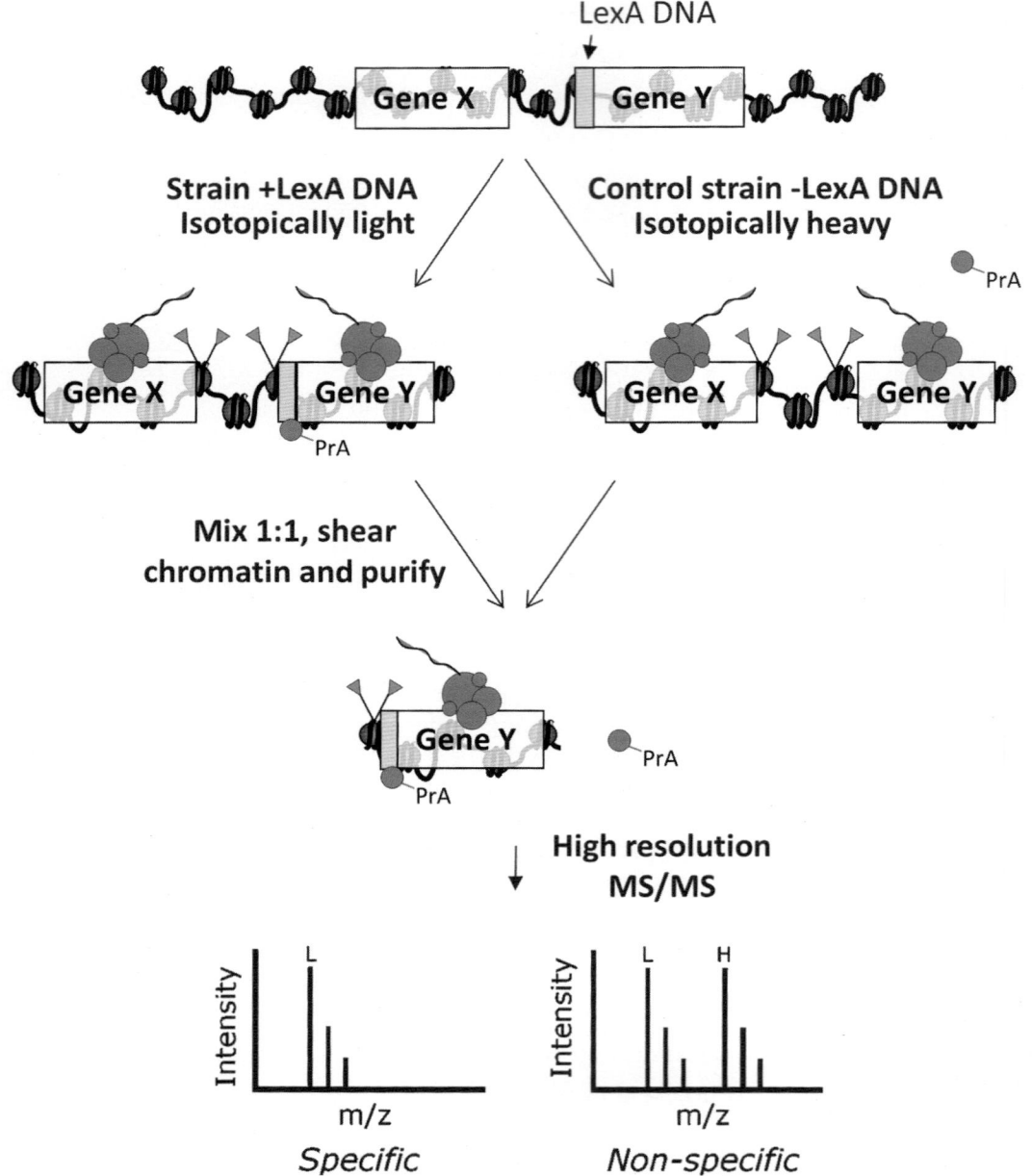

Fig. 1 Schematic representation of the ChAP-MS methodology. For the targeted purification of Gene Y for proteomic analysis, a LexA DNA binding site (*yellow*) is inserted near to the target gene by homologous recombination. A wild type strain without this LexA DNA binding site serves as a control. Each strain contains an ectopically expressed Protein A-tagged LexA fusion protein (*blue*). To label proteins, the strain with the LexA DNA binding site is grown in isotopically light medium while the control strain is cultured in isotopically heavy medium. Cross-linked cells from each culture are mixed, chromatin is sheared to ~1 kb in length by sonication, and Protein A-tagged LexA is collected on IgG-coated beads. Proteins and histones specifically enriched with the targeted Gene Y will be isotopically light as they arise only from the light culture. Proteins that are nonspecifically enriched in chromatin during purification will be a 1:1 mixture of isotopically light and heavy. High resolution mass spectrometry is used to determine the levels of isotopically light and heavy proteins, thus categorizing the enriched proteins and posttranslationally modified histones as either "specific" or contaminant (Color figure online)

from a plasmid grown in isotopically light media and (2) a strain without the LexA DNA binding site also expressing LexA-Protein A grown in isotopically heavy media. Both strains are grown in the absence of tryptophan to select for the *pLexA-Protein A* (*TRP1*) plasmid. The two strains are grown to equivalent densities, subjected to in vivo cross-linking with formaldehyde, and frozen independently. The cross-linking helps to stabilize the protein interactions and still allows for purification of native chromatin [9]. The frozen cells are mixed 1:1 prior to cryolysis with a ball mill maintained at liquid nitrogen temperature, which provides a method for generating a cell lysate without thawing. The ability to differentiate "specific" and "nonspecific" protein interactions with the affinity-tagged chromatin occurs at the point of thawing the cell lysate and extends for the duration of the purification procedure. During the course of the purification, specific protein interactions (which are exclusively isotopically light) with the affinity-tagged complex are maintained, while nonspecific protein associations have a 1:1 likelihood to occur from either the light or heavy proteins. The readout for these stable and nonspecific protein interactions with the affinity-tagged chromatin region is high resolution mass spectrometry. When peptides are assigned to a given protein co-purifying with the affinity-tagged chromatin, the type of protein interaction can be classified as specific if the peptides are ~100 % isotopically light or nonspecific if the peptides are ~50 % isotopically light. For simplicity, we detail below the overall approach used for ChAP-MS analysis of chromatin at the 5′ end of the *GAL1* gene in *S. cerevisiae* in transcriptionally active (+galactose) and inactive (+glucose) chromatin states [8].

The *LEXA::GAL1 pLexA-Protein A* strain was designed to have a LexA DNA binding site upstream of the *GAL1* start codon in the *S. cerevisiae* W303a background. First, the genomic *GAL1* gene was replaced with *URA3* using homologous recombination. Separately, the *GAL1* gene (+50 base pairs upstream and downstream) was amplified from genomic DNA by PCR with primers that incorporate a LexA DNA binding site (5′-CACTTGATACTGTATGAGCATAC AGTATAATTGC) immediately upstream of the *GAL1* start codon. The *LEXA::GAL1* PCR product was transformed into the *gal1::URA3* strain and selected for growth with 5-fluoroorotic acid, which is lethal in *URA3*-expressing cells. Positive transformants were sequenced to ensure homologous recombination of the cassette to create the *LEXA::GAL1* strain.

A plasmid that constitutively expresses LexA-Protein A fusion protein with *TRP1* selection was created by amplification of the Protein A sequence template *pOM60* via PCR and subcloning this into the SacI/SmaI ends of the expression plasmid *pLexA-C*. Transforming this plasmid into the *LEXA::GAL1* strain gave rise to the *LEXA::GAL1 pLexA-Protein A* strain. A control strain was constructed by transforming the *pLexA-Protein A* protein fusion plasmid into W303a *S. cerevisiae* (*see* **Note 1**).

3.1 Cell Cultures and Formaldehyde Treatment

1. Grow *S. cerevisiae LEXA::GAL1 pLexA-Protein A* and *pLexA-Protein A* strains in yeast synthetic media lacking tryptophan to mid-log phase (~3.0×10^7 cells/mL) at 30 °C with shaking at 200 rpm. The *LEXA::GAL1 pLexA-Protein A* strain is grown with isotopically light lysine, while the strain *pLexA-Protein A* is cultured exclusively with isotopically heavy $^{13}C_6{}^{15}N_2$-lysine [9]. Inoculate large-scale cultures (12 L) from 3 mL of stationary phase cultures in the respective synthetic medium, and grow in medium with either 2 % glucose or 3 % galactose with an estimated doubling time of 2.5 h to give ~5×10^{11} total cells per growth condition.

2. When the cultures reach ~3.0×10^7 cells/mL, remove the flasks from the incubator and add formaldehyde solution (37 % w/v) to a final concentration of 1.25 % (*see* **Note 2**). Swirl the flasks to mix and leave for 5 min at room temperature.

3. Add 2.5 M glycine solution to a final concentration of 125 mM glycine to quench the cross-linking. Swirl the flasks to mix and leave for 5 min at room temperature.

4. Collect the cells by centrifugation at 2,500 ×g in 1 L bottles at 4 °C for 30 min, wash with 100 mL of ice cold distilled H_2O, and re-collect by centrifugation. Add 20 mM Hepes, pH 7.5,1.2 % w/v polyvinylpyrrolidone solution to the wet cell pellet (1 mL/10 g of wet cells) and mix by pipetting.

3.2 Cryogenic Lysis

1. Fill a 50 mL conical polypropylene tube with liquid nitrogen. Slowly pipette the cell suspension drop-wise into the liquid nitrogen, using a 1 mL micropipette tip cut with scissors to remove ~1 cm from the end to provide easier pipetting. Add liquid nitrogen at intervals during the procedure to keep the tube full.

2. Once the cells are frozen as pellets, pour off the excess liquid nitrogen. Pierce three holes in the cap of the tube with a needle, place it on the tube, and store the frozen cells at –80 °C.

3. Weigh each set of pellets and mix the isotopically light *LEXA::GAL1 pLexA-Protein A* cells and isotopically heavy *pLexA-Protein A* cells 1:1 by weight in a conical 50 mL polypropylene tube. Shake the tube thoroughly to ensure mixing of the pellets. Keep the cells at liquid nitrogen temperature as much as possible during this process to avoid thawing.

4. Precool cylinders and ball bearings for the Retsch mixer mill in liquid nitrogen until the nitrogen stops boiling. Retrieve the cylinders from the liquid nitrogen using tongs. Perform cryogenic lysis using Retsch stainless steel cylinders. Twenty-five milliliter screw top grinding jars, and 20 mm stainless steel ball bearings: add ~3 g of mixed cell pellets to a cylinder with

a ball bearing and place it into liquid nitrogen. Once the cylinder is cooled, attach it to the mixer mill, process for 3 min at 30 Hz, and return it to liquid nitrogen. Repeat this cycle five times. Following the final cycle, open the cylinder and scoop out the cell powder into a 50 mL conical polypropylene tube placed in a bath of liquid nitrogen. Seal the tube with a cap containing three holes made with a needle and store at −80 °C (*see* **Note 3**).

3.3 Affinity Purification

1. Coat M270-epoxy Dynabeads with rabbit IgG for affinity purification via the Protein A-tag on LexA (*see* **Note 4**) by mixing 80 mg of M270-epoxy Dynabeads with 6 mg of rabbit IgG, 1 M ammonium sulfate, 60 mM sodium phosphate, pH 7.4 in a final volume of 5.36 mL and incubating overnight at 30 °C with rocking.

2. Collect the beads with a magnet and wash them successively with 1 mL of 100 mM glycine, pH 2.5; 1 mL of 10 mM Tris, pH 8.8; 1 mL of PBS four times; 1 mL of PBS/0.5 % Triton X-100; and 1 mL of PBS. All washes are done rapidly except for PBS/Triton X-100, which should be for 15 min.

3. All steps are performed at 4 °C. For a typical purification, we add 50 mL of affinity purification buffer to 10 g of cell powder containing isotopically light *LEXA::GAL1 pLexA-Protein A* and isotopically heavy *pLexA-Protein A* cell lysates.

4. Suspend the mixture by gentle inversion and blend thoroughly with a handheld Tissuemiser (or any available probe tissue homogenizer) for 20 s at 30,000 rpm.

5. Shear the chromatin in 20 mL aliquots by sonication with a Diagenode Bioruptor UCD-200 using the "low" power setting and 30 s on/off cycles for 12 min total time to yield ~1 kb chromatin fragments (*see* **Note 5**).

6. Centrifuge at 2,500×g for 10 min and collect the supernatant.

7. Add 80 mg of IgG-coated Dynabeads to the supernatant and invert gently on a rocking platform for ~4 h (*see* **Note 6**).

8. Collect the beads with a magnet and wash them five times with 1 mL of affinity purification buffer (without protease inhibitors) and three times with wash buffer.

9. Elute bound proteins from the beads in 0.5 mL of 0.5 N ammonium hydroxide/0.5 mM EDTA for 5 min at room temperature (*see* **Note 7**). Remove the beads with a magnet and lyophilize the eluate in a SpeedVac.

10. Resuspend the lyophilized proteins in 20 µL of SDS-PAGE gel loading buffer and heat at 95 °C for 20 min.

11. Resolve the eluted proteins in a 4–20 % gradient SDS-PAGE gel and stain the gel with GelCode Blue prior to imaging for documentation (*see* **Note 8**). All handling of the gel should be done with powder-free gloves and clean labware.

12. Place the gel on a clean glass plate with a ruler underneath and excise 2 mm-wide sections using a clean razor blade. Place each excised gel band into a 1.7 mL microcentrifuge tube and store at –20 °C.

13. Submit the gel bands to a proteomics facility for high resolution analysis of tryptic peptides and identification of proteins by mass spectrometry (*see* **Note 9**).

4 Notes

1. The strains of *S. cerevisiae* used in this protocol can be obtained from the authors upon request.

2. Several concentrations of formaldehyde cross-linking were tested to determine the amount needed to both trap protein-protein interactions and allow the chromatin to be soluble [10].

3. Cryogenic lysis with a mixer mill is the preferred method for lysing and blending the cells. One should avoid methods such as lysis with glass beads during which the samples will thaw, which precludes uniform blending prior to thawing. If a mixer mill is not available, a reasonable alternative is manual grinding with a mortar and pestle; the cells should be covered with liquid nitrogen during lysis and ground to a fine powder. Grinding should continue until >75 % lysis is observed using an optical microscope. Allow the liquid nitrogen to evaporate and store the ground cells at –80 °C.

4. Beads should be coupled freshly for each chromatin purification.

5. The number of cycles and duration of sonication for shearing chromatin to ~1,000 bp will vary between sonicators, and thus the parameters should be determined empirically.

6. Four mg of coupled Dynabeads are used for each gram of cell powder. The time of incubation has been determined empirically; shorter times (e.g., 1–4 h) can be tested and may result in fewer nonspecific interactions.

7. This elution procedure minimizes the amount of antibody heavy and light chains that are released from the resin. The affinity purification buffer described in this section has been determined empirically to provide good yields of

chromatin-associated protein complexes. Its components, e.g., NaCl and Triton X-100, can be varied according to the protein complex under study.

8. In our laboratory, SDS-PAGE is performed with pre-cast Invitrogen Tris–glycine gels. GelCode Blue, a colloidal stain, is recommended as it is sensitive for imaging purposes and easily removed during processing for mass spectrometry. To minimize keratin contamination, the major source of contamination during processing, cleaning items with a commercial glass cleaner such as Windex is recommended.

9. Proteomics facilities which are available at many universities and also commercially perform the following: destaining of the gel bands, in-gel trypsin digestion and tandem mass spectrometric identification of the tryptic peptides. A high resolution mass analyzer should be used to collect the mass spectra. Tandem mass spectra do not necessarily need to be collected with a high resolution mass analyzer. Proteins and posttranslationally modified histones will be identified by database searching of the tandem mass spectrometric data. Peak areas of peptides corresponding to the assigned proteins will be extracted from the mass spectra. Note that peak areas only need to be extracted for peptides that will contain the heavy amino acid(s). The percent of isotopically light peptide is calculated as $[(\text{light area}/(\text{heavy area}+\text{light area})]\times 100$. Percent light peptide values are averaged together for a given protein and the standard deviation is calculated. A typical representation of this data is a bar graph with percent light peptide values on the y-axis and the given protein on the x-axis [8] (Fig. 1). Specific protein interactions are seen as ~100 % isotopically light, while nonspecific associations are seen as ~50 % isotopically light. After identification of specific and nonspecific interactions, the investigator must select an appropriate system to explore the functional significance of the identified associating proteins. Knocking out or knocking down a particular protein provide good methods for studying the significance of the protein's interactions. qPCR-ChIP assays can also be performed to validate the proteomic findings [8].

Acknowledgments

We would like to acknowledge support for mass spectrometry from the UAMS Proteomics Facility and support from NIH grants R01GM106024, R33CA173264, UL1RR029884, P30GM103450, and P20GM103429.

References

1. Dedon PC, Soults JA, Allis CD et al (1991) Formaldehyde cross-linking and immunoprecipitation demonstrate developmental changes in H1 association with transcriptionally active genes. Mol Cell Biol 11:1729–1733

2. Ren B, Robert F, Wyrick JJ et al (2000) Genome-wide location and function of DNA binding proteins. Science 290:2306–2309

3. Pokholok DK, Harbison CT, Levine S et al (2005) Genome-wide map of nucleosome acetylation and methylation in yeast. Cell 122:517–527

4. Robertson G, Hirst M, Bainbridge M et al (2007) Genome-wide profiles of STAT1 DNA association using chromatin immunoprecipitation and massively parallel sequencing. Nat Methods 4:651–657

5. Johnson DS, Mortazavi A, Myers RM et al (2007) Genome-wide mapping of in vivo protein-DNA interactions. Science 316:1497–1502

6. Barski A, Cuddapah S, Cui K et al (2007) High-resolution profiling of histone methylations in the human genome. Cell 129:823–837

7. Mikkelsen TS, Ku M, Jaffe DB et al (2007) Genome-wide maps of chromatin state in pluripotent and lineage-committed cells. Nature 448:553–560

8. Byrum S, Raman A, Taverna SD et al (2012) ChAP-MS: a method for identification of proteins and histone posttranslational modifications at a single genomic locus. Cell Rep 2:198–205

9. Byrum SD, Smart SK, Larson S et al (2012) Analysis of stable and transient protein-protein interactions. Methods Mol Biol 833:143–152

10. Byrum SD, Taverna SD, Tackett AJ (2011) Quantitative analysis of histone exchange for transcriptionally active chromatin. J Clin Bioinform 1:17

Chapter 9

Chromatin Structure Analysis of Single Gene Molecules by Psoralen Cross-Linking and Electron Microscopy

Christopher R. Brown, Julian A. Eskin, Stephan Hamperl, Joachim Griesenbeck, Melissa S. Jurica, and Hinrich Boeger

Abstract

Nucleosomes occupy a central role in regulating eukaryotic gene expression by blocking access of transcription factors to their target sites on chromosomal DNA. Analysis of chromatin structure and function has mostly been performed by probing DNA accessibility with endonucleases. Such experiments average over large numbers of molecules of the same gene, and more recently, over entire genomes. However, both digestion and averaging erase the structural variation between molecules indicative of dynamic behavior, which must be reconstructed for any theory of regulation. Solution of this problem requires the structural analysis of single gene molecules. In this chapter, we describe a method by which single gene molecules are purified from the yeast *Saccharomyces cerevisiae* and cross-linked with psoralen, allowing the determination of nucleosome configurations by transmission electron microscopy. We also provide custom analysis software that semi-automates the analysis of micrograph data. This single-gene technique enables detailed examination of chromatin structure at any genomic locus in yeast.

Key words Chromatin, Site-specific recombination, Affinity purification, Psoralen cross-linking, Electron microscopy

1 Introduction

Nucleosomes are the basic subunits of chromatin and are ubiquitous throughout eukaryotic genomes, tightly regulating access to the genetic information encoded in DNA [1, 2]. The nucleosome core particle consists of an octameric protein complex, including two copies each of histones H3, H4, H2A, and H2B, and 147 base pairs of DNA wrapped around the histone core [3]. Histone variants and myriad posttranslational histone modifications generate structural and functional nucleosome diversity, which are thought to determine the accessibility and, therefore, the expression of individual genes. Recent high-throughput analyses of chromatin structure have generated genome-wide localization maps of nucleosomes

Ronald Hancock (ed.), *The Nucleus*, Methods in Molecular Biology, vol. 1228,
DOI 10.1007/978-1-4939-1680-1_9, © Springer Science+Business Media New York 2015

and histone modifications. However, these analyses of in vivo chromatin structure largely use variations of one method, the endonucleolytic digestion of DNA in isolated nuclei. Datasets of such analyses are derived from a large number of cells, in the range of thousands to millions [4]. Consequently, the chromatin structure at any particular genomic locus represents a population average. However, if the significance of chromatin structure for transcription is to be understood, then knowledge of its dynamic behavior is critical; this requires analysis of chromatin structure at the level of single gene molecules, rather than populations.

Recent quantitative models of gene expression have invoked promoter nucleosome remodeling as an important step in transcriptional activation. For example, models of transcriptional bursting assume random jumping of gene promoters between transcriptionally active (ON) and inactive (OFF) states. Random promoter nucleosome disassembly and reassembly events over DNA elements necessary for transcription have been proposed to underlie random ON/OFF transitioning [5, 6]. Initial attempts to study promoter nucleosome configurations at the *PHO5* gene supported the notion that multiple nucleosome configurations coexisted in a cell population [7, 8]. However, theoretical alternatives could not be refuted without "direct" demonstration that alternative nucleosome configurations indeed existed. Furthermore, without knowledge of the relative frequency distribution of nucleosome configurations, the theory of random transitioning between alternative chromatin structures remained untestable.

To overcome these limitations, we developed a method that allows for the determination of promoter nucleosome configurations by electron microscopy on single gene molecules in the yeast *Saccharomyces cerevisiae* [9]. The technique involves flanking a genomic region of interest with recombination elements from the yeast *Zygosaccharomyces rouxii*, which enables the excision of the region in the form of a chromatin ring [10–13]. Rings are purified via a cluster of LexA binding sites that is recognized by a recombinant LexA protein fused to a tandem affinity purification (TAP) tag on its C-terminus. Purified rings are cross-linked with psoralen, a DNA intercalating agent that covalently cross-links pyrimidines on opposite strands of the DNA double helix upon exposure to long-wave UV light [14–19]. Importantly, nucleosome-wrapped DNA is protected from psoralen cross-linking while linker DNA between nucleosomes is cross-linked efficiently [20]. The single gene molecules are denatured prior to electron microscopy, forming single-strand DNA "bubbles" at positions previously occupied by nucleosomes. These single-strand bubbles are visible in the electron microscope; highly cross-linked linker DNA remains double-stranded due to the covalent interstrand psoralen cross-links (Fig. 1). A variation of this cross-linking technique was used previously to distinguish between transcriptionally active open and nucleosomal closed yeast rDNA chromatin states and to characterize the chromatin structure of SV40 mini-chromosomes [21–23].

In this chapter, in addition to providing a detailed description of the purification and psoralen cross-linking of single gene molecules we make available a custom-written single-molecule analysis program, created using MATLAB code, for general use. The method described herein has been used to investigate the stochastic nature of *PHO5* promoter chromatin remodeling [9] and to study nucleosome configurations at the yeast 5S rDNA locus [24]. Importantly, this method is amenable for the study of any gene or locus of interest in yeast, and promises to provide additional insight into the complex nature of chromatin remodeling and its role in transcriptional regulation and gene expression.

2 Materials

2.1 Strain Construction, Culturing, and Harvesting

1. A *Saccharomyces cerevisiae* strain with the gene or region of interest flanked by recombination sites (RS elements) for site-specific recombination and containing an internal cluster of LexA binding sites for affinity purification (*see* **Note 1**).

2. Synthetic Complete dropout medium lacking leucine (SC-L), supplemented with either 2 % glucose (SCD-L) or 2 % raffinose (SCR-L).

3. Plasmid pSH17 (K2049) [9, 24] for constitutive expression of the LexA-TAP fusion protein (under control of a *TEF2* promoter) and galactose-inducible expression of the R recombinase (under control of a *GAL1* promoter) (*see* **Note 1**).

4. Galactose: 40 % (w/v) solution in distilled water.

5. 10 L-capacity fermentor with temperature, agitation, and airflow control (New Brunswick Scientific or similar).

6. 20 mL plastic syringes.

7. Plastic Tube Topper (Beckman Coulter).

8. Liquid N_2.

2.2 Affinity Purification of Chromatin Rings

2.2.1 Cell Lysis and Differential Centrifugation

1. Commercial coffee grinder (Tefal Prepline or similar).

2. Dry ice pellets: 160–220 g per preparation.

3. Protease inhibitors (PIs, 200×): 0.4 mM pepstatin A, 0.4 mM E-64 in DMSO, store in small aliquots at –20 °C.

4. 4-(2-Aminoethyl) benzenesulfonyl fluoride hydrochloride, AEBSF (500×): 1.4 mM in 10 mM NaOAc, pH 5.2, store at 4 °C.

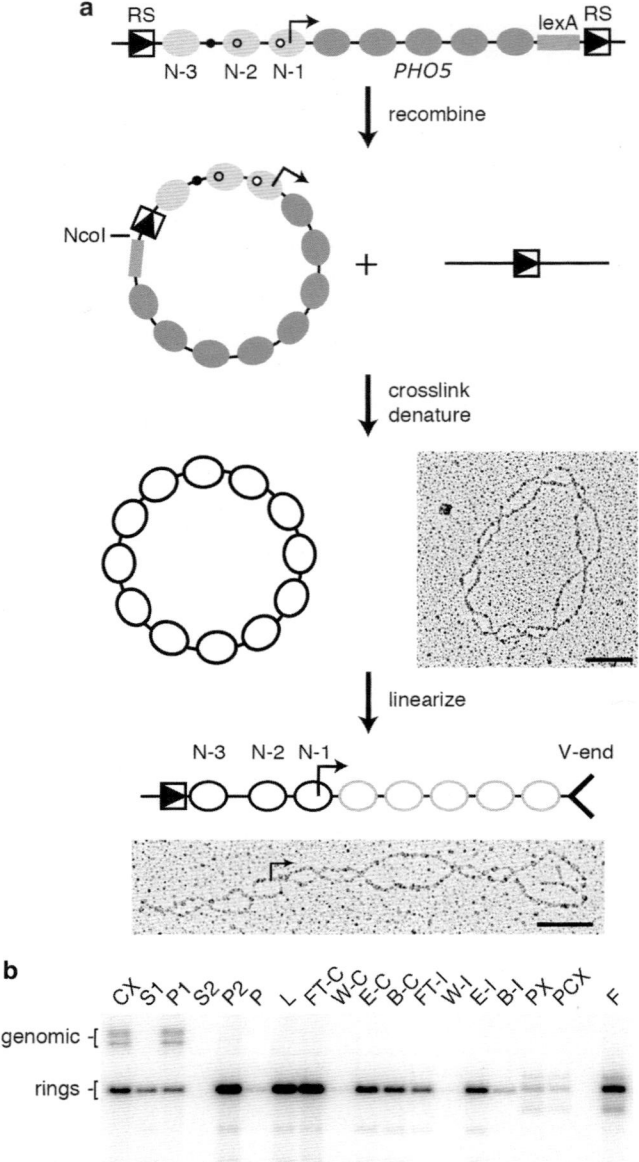

Fig. 1 Single gene molecule purification and psoralen cross-linking. (**a**) Recombination elements (RS) from *Zygosaccharomyces rouxii* are incorporated upstream and downstream of a genomic locus of interest in *Saccharomyces cerevisiae*, in this case the *PHO5* gene [9–11]. Upon expression of a site-specific recombinase, the locus is excised from the genome as an extra-chromosomal gene ring [12]. A constitutively expressed fusion protein containing LexA fused to a tandem-affinity purification (TAP) tag binds to a lexA operator cluster located next to the 3′ RS element. Purified gene rings are cross-linked with psoralen and purified. DNA is denatured and prepared for electron microscopy. Gene rings contain "bubbles" of single-stranded DNA, indicating positions previously

5. Extraction Buffer (XB): 25 mM HEPES, pH 7.4, 200 mM KOAc, 2 mM MgCl$_2$, 10 % glycerol (v/v), 0.125 mM spermidine, 0.05 mM spermine, 1× PIs, 1× AEBSF, 5 mM 2-mercaptoethanol. Prepare freshly and cool to 4 °C prior to use.

6. 2-mercaptoethanol (14.3 M).

7. Ultracentrifuge: Optima L-90K with 45 Ti and 70 Ti rotors (Beckman Coulter) or similar.

8. Stirring hotplate: Cimarec analog basic (Thermo Fisher) or similar.

2.2.2 Tandem Affinity Chromatography: Calmodulin Resin Columns

1. PolyPrep columns (Bio-Rad).

2. Calmodulin affinity resin (Agilent Technologies).

3. Calmodulin Binding Buffer (CBB): 25 mM HEPES, pH 7.4, 200 mM KOAc, 1 mM MgCl$_2$, 1 mM imidazole, 2 mM CaCl$_2$, 0.01 % (v/v) NP-40. Store at 4 °C. Prior to use, add protease inhibitors (PIs and AEBSF, final concentration 1×) and 2-mercaptoethanol (final concentration 5 mM).

4. Calmodulin Wash Buffer (CWB): 25 mM HEPES, pH 7.4, 400 mM KOAc, 1 mM MgCl$_2$, 1 mM Imidazole, 2 mM CaCl$_2$, 0.1 % NP-40 (v/v). Store at 4 °C. Prior to use, add protease inhibitors (PIs and AEBSF, final concentration 1×) and 2-mercaptoethanol (final concentration 5 mM).

5. Calmodulin Elution Buffer (CEB): 25 mM HEPES, pH 7.4, 200 mM KOAc, 10 mM EGTA, 1 mM EDTA. Store at 4 °C. Prior to use, add protease inhibitors (PIs and AEBSF, final concentration 1×) and 2-mercaptoethanol (final concentration 5 mM).

6. StarHead magnetic stir bars (10 mm diameter × 8 mm height, Thermo Fisher).

7. Centrifuge: Avanti J-26 XPI with JLA-16.250 rotor and tube adapters cat. 356997 (Beckman Coulter) or similar.

Fig. 1 (continued) occupied by nucleosomes. Gene rings are linearized by restriction enzyme digestion with NcoI. Linearized molecules are oriented by a single-stranded DNA "V-end" at the 3′ end of the molecule, which arises from the LexA fusion protein protecting the lexA operator cluster from psoralen cross-linking. Scale bars 100 nm. (**b**) Southern blot analysis of a *PHO5* gene ring preparation. DNA from samples collected during the gene ring preparation were purified and resuspended in 100 μL TE. *PHO5* rings were linearized by digesting 10 μL of each sample with NcoI. Digests were loaded onto a 1 % agarose gel and hybridized with a *PHO5*-specific probe. Sample identities and amounts analyzed (in parentheses) on the gel: CX, cellular extract (0.017 %); S1, first supernatant (0.01 %); P1, first pellet (0.01 %); S2, second supernatant (0.01 %); P2, second pellet (0.063 %); P, pellet to remove insoluble material (0.063 %); L, calmodulin load (0.063 %); FT-C, calmodulin flow-through (0.125 %); W-C, calmodulin wash 0.05 %); E-C, calmodulin elution (0.167 %); B-C, calmodulin beads (0.167 %); FT-I, IgG flow-through (0.333 %); W-I, IgG wash (0.1 %); E-I, IgG elution (0.286 %); B-I, IgG beads (0.286 %); PX, post psoralen cross-linking (0.222 %); PCX, phenol–chloroform extraction (0.1 %); F, final sample ready for EM preparation (3.13 %). Positions of genomic DNA and ring DNA are indicated to the *left* of the blot

1. IgG-Sepharose 6 Fast Flow (GE Healthcare).

2. IgG Wash Buffer (IGW): 25 mM HEPES, pH 7.4, 200 mM KOAc, 1 mM EDTA, 0.005 % NP-40 (v/v). Store at 4 °C. For column equilibration, supplement with spermidine (0.125 mM) and spermine (0.05 mM). Following binding of samples to the column, use without supplements.

3. 0.5 M spermine.

4. 0.5 M spermidine.

5. IgG Cleavage Mix (ICM): 25 mM HEPES, pH 7.4, 200 mM KOAc, 1 mM EDTA. Store at 4 °C.

6. Tobacco Etch Virus (TEV) protease, 6× His-tagged (100 μg per preparation) (Sigma-Aldrich) or can be homemade [25].

7. NP-40.

2.3   Psoralen Cross-Linking

1. DNA control for cross-linking efficiency (different length from sample).

2. Psoralen solution: 4,5′,8-trimethylpsoralen (TMP) (Sigma-Aldrich), 1 mg/mL in 100 % ethanol, store in light-safe tubes at –20 °C. Bring to room temperature and vortex rigorously prior to use.

3. UV cross-linker: Stratalinker 2400 with five 366 nm 15-Watt bulbs.

4. 10 cm-diameter petri dishes.

5. IRN Buffer: 50 mM Tris–HCl, pH 8.0, 20 mM EDTA, 0.5 M NaCl.

6. RNase A (Sigma-Aldrich) 20 mg/mL in 10 mM Tris–HCl, pH 8.0, 50 % (v/v) glycerol. Store at –20 °C.

7. Proteinase K (Sigma-Aldrich) 20 mg/mL in 40 % (v/v) glycerol containing 10 mM Tris–HCl, pH 7.5, and 1 mM calcium acetate, store at 4 °C.

8. Phenol–chloroform–isoamyl alcohol (25:24:1, Thermo Fisher), store at 4 °C in a light-safe bottle.

9. Glycogen (20 mg/mL aqueous solution) (Roche).

10. 5 M NaCl.

11. 100 % ethanol.

12. Polyallomer centrifuge tubes (38.5 mL, cat. 355642, Beckman Coulter).

13. 10 mM Tris–HCl, pH 8.

14. NcoI and 10× NE Buffer 3 (New England Biolabs).

15. SafeSeal microcentrifuge tubes, 1.7 mL, low-binding (Sorenson BioScience).

16. DNA Clean and Concentrator-5 Kit (Zymo Research).

17. TEN Buffer: 30 mM tetraethylammonium chloride, pH 7.8, 20 mM EDTA, 10 mM NaCl.

2.4 DNA and Protein Analysis

1. TE Buffer: 10 mM Tris–HCl, pH 7.5, 1 mM EDTA.

2.4.1 DNA Analysis

2.4.2 Protein Analysis

1. SDS-PAGE sample buffer (3×): 150 mM Tris–HCl, pH 6.8, 6 % (w/v) SDS, 30 % (v/v) glycerol, 3 % (v/v) 2-mercaptoethanol, 0.06 % (w/v) bromophenol blue.

2. SDS-PAGE gels: 10 %, made using standard techniques.

3. Goat anti-CBP antibody (W-15) (Santa Cruz sc-32998).

4. Donkey anti-goat IgG, horseradish peroxidase (HRP) conjugated (Santa Cruz sc-2020).

2.5 Preparation of DNA for Electron Microscopy

2.5.1 Preparation of Ethidium Bromide Carbon (EBC) Grids

1. Metal evaporator and thickness monitor (Cressington Scientific Instruments, Watford, UK).

2. Mechanical Ultrasonic Cleaner water bath (0.5 L, Thermo Fisher).

3. Acetone (Thermo Fisher).

4. Carbon rods (Electron Microscopy Sciences).

5. Ethidium bromide solution: 30 μg/mL in filtered water, made fresh.

6. Copper grids, 400 lines/in., square mesh (G400-Cu, Electron Microscopy Sciences).

7. Muscovite Mica, V-5 (Electron Microscopy Sciences).

8. Dumont N-5 tweezers (Electron Microscopy Sciences).

9. Whatman filter paper.

10. Double-deionized water (Milli-Q).

11. Smith Grid Coating Trough (Ladd Research, Williston, VT, USA).

2.5.2 Denaturing DNA Samples

1. Benzalkonium chloride (BAC): 0.2 % (w/v) in deionized formamide, stable at room temperature for 1 year (Sigma-Aldrich).

2. Formamide: 99 %, deionized (Sigma-Aldrich).

3. Glyoxal: 8.8 M (Sigma-Aldrich).

2.5.3 Spreading Denatured DNA Samples

1. 5- and 10-cm-diameter petri dishes.

2. Graphite powder (Thermo Fisher).

3. Uranyl acetate: 2 % (w/v) solution in Milli-Q water.

4. EBC grids: made fresh, use within 2 h.

5. Grid storage boxes (Electron Microscopy Sciences).

2.5.4 Rotary Metal Shadowing

1. Metal evaporator, rotary stage, and thickness monitor (Cressington Scientific Instruments).

2. Lint-free Kimwipes (Kimberly-Clark).

3. Glass microscope slides.

4. Double-sided scotch tape.

5. Platinum–palladium (80:20) wire, 0.2 mm diameter, 99.95 % (Electron Microscopy Sciences).

6. Tungsten wire baskets, 3-strand (Ted Pella, Redding, CA, USA).

2.6 Electron Microscopy

1. Transmission electron microscope: 120 keV (JEOL) or equivalent

2. Digital CCD camera (4K×4K resolution, Gatan).

2.7 Single Gene Molecule Analysis

2.7.1 Tracing Single Gene Molecules

1. ImageJ software: available free from http://rsbweb.nih.gov/ij/download.html).

2. Touchscreen tablet computer (iPad, Apple).

3. Screen mirroring software to run ImageJ on the tablet (Air Display, Apple).

4. Touchscreen fine-tip stylus (JotPro, Adonit).

2.7.2 MATLAB Analysis of Single Gene Molecules

1. Custom program written in MATLAB: download code and more detailed instructions at http://bio.research.ucsc.edu/people/boeger/EM_Analysis_Software.htm.

3 Methods

3.1 Strain Construction, Culturing, and Harvesting

1. Integration of recombination elements (RS) and LexA binding sites into a genomic region of interest in *Saccharomyces cerevisiae* has been described in detail [10, 11] (*see* Fig. 1a for a schematic representation of the genetically modified *PHO5* gene locus).

2. Yeast strains are transformed with plasmid pSH17 (*LEU2* selection) to introduce the constitutively expressed LexA-TAP fusion protein under the control of the *TEF2* promoter and the galactose-inducible R recombinase gene. Transformed strains are grown on SCD-L or SCR-L as selection for plasmid maintenance. The *LEU2* selection gene is flanked by RS elements that excise it under conditions of inappropriate R recombinase expression (prior to galactose induction) (*see* **Note 2**).

3. Yeast pre-cultures are grown to saturation in 100 mL of SCD-L at 30 °C.

4. 10 L of SCR-L are sterilized in a fermentor and cooled to 30 °C.

5. The fermentor culture is inoculated at 2×10^5 cells/mL and grown overnight at 30 °C with agitation set to 250 rpm and airflow set to 4 Standard Liters per Minute (SLPM) (*see* **Note 3**).

6. Upon reaching a target density of 4–5×10^7 cells/mL (4–5×10^{11} cells total), galactose is added to a final concentration

of 2 % to induce the R recombinase and excision of gene rings (Fig. 1a).

7. Galactose induction is continued for 90 min during which time the cells continue to grow at 30 °C with agitation (250 rpm) and airflow (4 SLPM). Approximately 75 % of RS-flanked loci are excised as chromatin rings during this incubation [10, 11].

8. Cells are harvested by centrifugation (5 min, 9,000×g, room temperature) yielding 3–4 g of wet cells/L of culture (30–40 g total).

9. Cells are washed twice with sterile water before pelleting in sealed 20 mL syringes (5 min, 6,000×g, 4 °C) (*see* **Note 4**).

10. Supernatants are discarded and the syringe tips are cut open with a scalpel. A plunger is placed in the syringe and firmly pressed down, extruding the yeast into a 200 mL plastic beaker filled with liquid N2.

11. This frozen yeast "spaghetti" is broken into smaller clumps with one of the empty syringes, quickly drained of liquid N2, and poured into 50 mL conical tubes. The gram weight of the frozen yeast cells is measured quickly prior to storing the tubes at –80 °C. Yeast spaghetti may be stored indefinitely at –80 °C.

3.2 Affinity Purification of Chromatin Rings

3.2.1 Cell Lysis and Differential Centrifugation

1. All steps should be carried out at 4 °C unless otherwise noted.

2. The coffee grinder is precooled by grinding two 50 g batches of dry ice for 30 s. The powdered dry ice from these two steps is discarded.

3. 60–120 g of dry ice (3× the weight of the cells to be processed) is placed in the grinder and ground for 30 s. The resulting dry ice powder is kept in the grinder.

4. 20–40 g of frozen yeast are placed on top of the dry ice powder.

5. The grinder's lid is sealed with Parafilm and the yeast are ground, with rigorous shaking, for 4 × 1 min with 30 s breaks to prevent overheating (*see* **Note 5**).

6. The dry ice/yeast powder is placed in a 1 L plastic bucket containing a large magnetic stir bar.

7. The dry ice is evaporated with constant, gentle stirring on a magnetic stir plate for 15 min. The frozen yeast powder may be stored at –80 °C.

8. Four volumes (4× the gram weight of the spaghetti) of precooled Extraction Buffer (XB) is added to the powder and stirred for 30 min (*see* **Note 6**).

9. Samples of the cell extract (CX) are taken for DNA (200 μL) and protein (20 μL) analyses and kept on ice until storage at –80 °C.

10. The cell extract is distributed into four 45 Ti-compatible ultracentrifuge tubes and balanced.

11. Cell debris is pelleted by centrifugation in an Optima L-90K Ultracentrifuge (1 h, 45 Ti rotor, $73,000 \times g$, 4 °C).

12. During this centrifugation a 250 mL glass graduated cylinder is cooled to 4 °C, and the coffee grinder is cleaned with water and 70 % ethanol.

13. Following centrifugation the supernatants (S1), which contain the chromatin rings, are combined in the precooled cylinder and brought to 200 mL with buffer XB. The cylinder is sealed with Parafilm and mixed by gentle inversion. The pellet of cell debris (P1) is set aside until further processing (**step 18**).

14. Samples of the S1 supernatant are taken for DNA (200 μL) and protein (20 μL) analyses and kept on ice until storage at −80 °C.

15. 25 mL aliquots of solution S1 are pipetted into eight 70 Ti ultracentrifuge tubes. The tubes are balanced in pairs to ≤0.1 g before loading into a precooled 70 Ti rotor.

16. Chromatin rings are pelleted by centrifugation in an Optima L-90K ultracentrifuge (2.5 h, $371,000 \times g$, 4 °C) (*see* **Note 7**).

17. During centrifugation, the P1 pellets of cell debris are suspended in buffer XB to a final volume of 200 mL.

18. Samples of the P1 pellet suspension are taken for DNA (200 μL) and protein (20 μL) analyses and kept on ice until storage at −80 °C. The remaining P1 suspension is discarded.

19. Columns for tandem affinity purification are equilibrated during this centrifugation (*see* below).

3.2.2 Tandem Affinity Chromatography: Calmodulin Resin Columns

1. All steps should be carried out at 4 °C unless otherwise noted.

2. Four PolyPrep columns are loaded with 750 μL of calmodulin resin. Two additional columns are loaded with 400 μL of IgG-sepharose beads. The storage buffer is allowed to drain by gravity flow.

3. The calmodulin columns are equilibrated with 2×10 mL of Calmodulin Binding Buffer (CBB) without protease inhibitors or 2-mercaptoethanol. The IgG-sepharose columns are equilibrated with 2×10 mL of IgG Wash Buffer (IWB) containing 0.125 mM spermidine and 0.05 mM spermine.

4. Approximately 5 mL of buffer CBB and buffer IWB from the second wash steps are kept in the column by sealing the bottoms with the stoppers and the tops with the caps provided with the columns (*see* **Note 8**).

5. The calmodulin resin columns are rotated on a wheel for 2 h at 4 °C, while the IgG-sepharose columns are rotated for 3 h at 4 °C to equilibrate the resin/beads with the binding buffers.

6. Near the end of the centrifugation of the S1 solution, 70 mL of buffer CBB are supplemented with 1× PIs, 1× AEBSF, and 5 mM 2-mercaptoethanol.

7. A Beckman Avanti J-26 XPI centrifuge with a JLA-16.250 rotor containing adaptors for 2 × 50 mL conical tubes is precooled to 4 °C.

8. Following centrifugation of the S1 solution, the second supernatant (S2) from the eight centrifuge tubes is pooled into a precooled glass beaker.

9. Samples of S2 are taken for DNA (200 µL) and protein (20 µL) analyses and kept on ice until storage at −80 °C. The remaining S2 is discarded.

10. Each pellet from the second spin (P2) is washed quickly with 3 × 1 mL of buffer CBB (including 1× PIs, 1× AEBSF, and 5 mM 2-mercaptoethanol) to remove residual glycerol.

11. Residual buffer CBB is removed from the pellets before adding an additional 1 mL of buffer CBB (including 1× PIs, 1× AEBSF, and 2-mercaptoethanol) to each pellet.

12. The sticky pellets are detached from the sides of the tubes with a 200 µL pipette tip by scraping up and down. The tips are finally ejected into the tubes, because the pellets will be partially stuck to them and a significant amount of material would be lost should they be removed.

13. Small StarHead magnetic stir bars are added to the tubes, which are then placed vertically next to each other in a plastic beaker. The beaker is placed on a magnetic stir plate and set to gentle agitation for 20 min to allow for pellet resuspension.

14. A further 1 mL of buffer CBB (including 1× PIs, 1× AEBSF, and 2-mercaptoethanol) is added to each tube and stirred for an additional 20 min. The solutions are pooled in a polyallomer centrifuge tube (16 mL total volume).

15. Samples of the P2 suspension are taken for DNA (100 µL) and protein (20 µL) analyses and kept on ice until storage at −80 °C.

16. The P2 suspension is centrifuged in the precooled JLA-16.250 rotor (5 min, 25,000 × g, 4 °C) to remove insoluble material that could block the chromatography columns.

17. While the P2 suspension is centrifuging, the four columns containing calmodulin resin are drained by gravity flow.

18. Following the P2 spin, the supernatant that will be loaded (L) onto the columns is placed in a new 15 mL conical on ice.

19. Samples of the load (L) are taken for DNA (100 µL) and protein (20 µL) analyses and kept on ice until storage at −80 °C.

20. Before continuing with affinity purification, the load (L) may be frozen and stored at −80 °C after adding glycerol to a final concentration of 10 % (v/v).

21. The insoluble pellet (P) from **step 16** is resuspended with 16 mL of buffer CBB (including 1× PIs, 1× AEBSF, and 5 mM 2-mercaptoethanol).

22. Samples of P are taken for DNA (100 µL) and protein (20 µL) analyses and kept on ice until storage at –80 °C. The remaining P is discarded.

23. The calmodulin columns are stoppered and 4 mL of load (L) are applied to each. The columns are capped and incubated on a rotator for 1 h at 4 °C. The columns are opened and the flow-through (FT-C) is collected and pooled.

24. 50 mL of Calmodulin Wash Buffer (CWB) and 25 mL of Calmodulin Elution Buffer (CEB) are supplemented with 1× PIs, 1× AEBSF, and 5 mM 2-mercaptoethanol and placed on ice.

25. Samples of FT-C are taken for DNA (200 µL) and protein (20 µL) analyses and kept on ice until storage at –80 °C. The remaining FT-C is discarded.

26. The columns are washed with 25× bead volumes (10 mL) of buffer CWB (including 1× PIs, 1× AEBSF, and 5 mM 2-mercaptoethanol). All washes (W-C) are pooled.

27. Samples of W-C are taken for DNA (200 µL) and protein (20 µL) analyses and kept on ice until storage at –80 °C. The remaining W-C is discarded.

28. A 15 mL conical tube is placed under each column. 500 µL of buffer CEB (including 1× PIs, 1× AEBSF, and 5 mM 2-mercaptoethanol) is added to each column and allowed to elute into the tubes by gravity flow.

29. The columns are stoppered, an additional 500 µL of buffer CEB (including 1× PIs, 1× AEBSF, and 5 mM 2-mercaptoethanol) is added to each, and the columns are capped and incubated on a rotator for 20 min at 4 °C.

30. The columns are opened and the eluates are collected in the 15 mL conical tubes.

31. A final 500 µl of buffer CEB (including 1× PIs, 1× AEBSF, and 5 mM 2-mercaptoethanol) is added to each column and allowed to elute into the 15 mL conical tubes.

32. Columns are placed in 50 mL conical tubes and centrifuged gently (30 s, $400 \times g$, at 4 °C) to collect residual eluate (~750 µL).

33. Eluates (E-C) are combined in a 15 mL conical tube, total volume 6 mL.

34. Samples of eluates E-C are taken for DNA (200 µL) and protein (20 µL) analyses and kept on ice until storage at –80 °C.

35. The columns are stoppered and 1 mL of buffer CEB (including 1× PIs, 1× AEBSF, and 5 mM 2-mercaptoethanol) is added

to each. The resin is resuspended in the buffer CEB, pooled into a 15 mL conical tube, and the total volume is brought up to 6 mL with buffer CEB.

36. Samples of the resin slurry (B-C) are taken for DNA (100 µL) and protein (20 µL) analyses and kept on ice until storage at −80 °C (*see* **Note 9**). The remaining B-C is discarded.

3.2.3 Tandem Affinity Chromatography: IgG-Sepharose Columns

1. All steps should be carried out at 4 °C unless otherwise noted.

2. The two IgG-sepharose columns are drained by gravity flow.

3. The columns are stoppered, 3 mL of the eluate (E-C) from the calmodulin columns is applied to each, and the columns are capped and incubated on a rotator for 3 h at 4 °C.

4. The columns are opened and the flow-through (FT-I) is collected and pooled.

5. Samples of FT-I are taken for DNA (200 µL) and protein (20 µL) analyses and kept on ice until storage at −80 °C. The remaining FT-I is discarded.

6. The columns are washed with 50× bead volumes (10 mL) of buffer IWB (without PIs, AEBSF, 2-mercaptoethanol, spermidine, or spermine). All washes (W-I) are pooled.

7. Samples of W-I are taken for DNA (200 µL) and protein (20 µL) analyses and kept on ice until storage at −80 °C. The remaining W-I is discarded.

8. The columns are stoppered and loaded with 3× bead volumes (600 µL) of IgG Cleavage Mix (ICM) to which TEV protease (50 µg) has been added.

9. The columns are capped and incubated on a rotator overnight (for at least 12 h) at 4 °C.

10. The following day, the columns are eluted by gravity flow and the eluate is collected in 15 mL conical tubes.

11. 1 mL of buffer ICB (containing 0.005 % v/v NP-40, is added to each column and eluted by gravity flow into the 15 mL conical tubes.

12. The columns are placed in 50 mL conical tubes and centrifuged gently (30 s, 400×*g*, at 4 °C) to collect residual eluate (~400 µl).

13. All the eluates (E-I) are combined in a 15 mL conical for a total volume of 3.5 mL.

14. Samples of E-I are taken for DNA (100 µL) and protein (20 µL) analyses and kept on ice until storage at −80 °C.

15. The columns are stoppered and 1 mL of buffer ICB (containing 0.005 % NP-40, v/v) is added. The beads are resuspended in the ICB and pooled into a 15 mL conical tube. The total volume is increased to 3.5 mL with buffer ICB (containing 0.005 % NP-40, v/v).

16. Samples of the IgG-sepharose bead slurry (B-I) are taken for DNA (100 μL) and protein (20 μL) analyses and kept on ice until storage at –80 °C (*see* **Note 9**). The remaining B-I is discarded.

17. The volume of the E-I sample is brought up to 4 mL with buffer ICB (containing 0.005 % NP-40, v/v) to facilitate spreading for psoralen cross-linking.

3.3 Psoralen Cross-Linking

1. Fifty nanograms of purified control DNA is added to the E-I sample to control for psoralen cross-linking efficiency (*see* **Note 10**).

2. Five 366 nm 15 W UV bulbs are mounted in a Stratalinker 2400 (*see* **Note 11**).

3. A working solution of 4,5′,8-trimethylpsoralen (TMP, psoralen) is made by diluting the 1 mg/mL stock solution (in 100 % ethanol) to 400 μg/mL in 100 % ethanol. Opaque tubes are used for the working solution and enough is made for 7×200 μL applications per sample to be cross-linked.

4. An ice–water slurry is made in a styrofoam container such that a 10 cm petri dish floating on top (without lid) is ~5 cm from the UV bulbs in the Stratalinker.

5. The E-I sample (4 mL) is added to the petri dish and swirled until it completely covers the surface.

6. 200 μL of 400 μg/ml TMP (0.05 volumes) is added and the petri dish is swirled gently to mix.

7. The petri dish is placed carefully on top of the ice–water slurry and placed in the Stratalinker. The door is closed and the sample is incubated in the dark for 5 min to allow TMP to incorporate into the DNA.

8. Cross-linking is performed for 5 min using the "Time" function on the Stratalinker.

9. **Steps 6–8** are repeated six more times (7 cycles total, 35 min of UV irradiation).

10. The sample (4.5 mL total) is removed from the petri dish and placed into a light-protected 50 mL conical tube.

11. Samples of the psoralen-cross-linked (PX) DNA are taken for DNA (100 μL) and protein (20 μL) analyses and kept on ice until storage at –80 °C.

12. The petri dish is washed with an equal volume (4.5 mL) of buffer IRN, which is added to the 50 mL conical tube (*see* **Note 12**).

13. RNase A (20 mg/mL) is added to a final concentration of 330 μg/mL and the sample is incubated for 2 h at 37 °C.

14. Proteinase K (20 mg/mL) is added to a final concentration of 330 μg/mL and 10 % SDS to a final concentration of 0.5 % (v/v). The sample is incubated for 4 h at 55 °C.

15. An equal volume (10 mL) of phenol–chloroform–isoamyl alcohol is added to the sample. The tube is shaken vigorously before placing it on a rotator at room temperature for 10 min. In addition to removing protein contaminants, this step removes excess soluble TMP from the solution, preventing further cross-linking.

16. The sample is centrifuged in a JLA-16.250 rotor with adaptors for 50 mL conical tubes (10 min, $5,000 \times g$, 4 °C).

17. Samples of the phenol–chloroform extraction (PCX) are taken for DNA (100 μL) and protein (20 μL) analyses and kept on ice until storage at –80 °C.

18. The aqueous (top) phase is divided into two open-topped 38.5 mL polyallomer tubes (~4.75 mL in each) and 2 volumes of 100 % ethanol are added, along with glycogen solution to a final concentration of 13 μg/mL and NaCl to a final concentration of 200 mM. The tubes are covered with Parafilm and inverted several times to mix the samples.

19. The DNA is pelleted by centrifugation in a JLA-16.250 rotor with adaptors (20 min, $21,000 \times g$, 4 °C).

20. The DNA pellets are washed with 10 mL of 70 % ethanol and the tubes are covered with Parafilm, inverted to mix, and centrifuged again (5 min, $21,000 \times g$, 4 °C).

21. The DNA pellets are dried for 10 min at room temperature, resuspended in a total volume of 100 μL with 10 mM Tris–HCl buffer, pH 8.0, and transferred to a low-binding centrifuge tube. The DNA may be stored at –80 °C.

22. The DNA is linearized by digesting with NcoI overnight at 37 °C in a low-binding centrifuge tube (*see* **Note 13**).

23. The digested DNA is column-purified using the Clean and Concentrator Kit and eluted with 8 μL of buffer TEN into a low-binding centrifuge tube. 0.25 μL of this final sample (F) is mixed with 9.75 μL of 10 mM Tris–HCl, pH 8.0 in a low-binding centrifuge tube and stored at –80 °C for use in Southern blot analysis.

24. Typical preparations yield 40–120 fmol of *PHO5* gene rings from 4×10^{11} cells, a recovery of ~6–18 % [10, 11]. The linearized DNA may be stored at –80 °C until preparation for EM (Subheading 3.5).

3.4 DNA and Protein Analysis

3.4.1 DNA Analysis

1. DNA collected during ring preparation is brought to 400 μL with buffer IRN. 4 μL of proteinase K (20 mg/mL) and 20 μL of SDS (10 % w/v) are added and the samples are incubated for 2 h at 55 °C.

2. DNA is extracted with phenol–chloroform–isoamyl alcohol (*see* Subheading 3.3, **step 15**).

3. 1 µL of RNAse A (20 mg/mL) is added and the samples are incubated for 1 h at 37 °C.

4. 2× volumes of 100 % ethanol are added, along with glycogen (20 mg/mL) to a final concentration of 13 ng/µL and NaCl to a final concentration of 200 mM. DNA is precipitated for 20 min at –20 °C.

5. The DNA is pelleted by centrifugation (30 min, $21,000 \times g$, 4 °C) and the pellets are washed with 500 µL of 70 % ethanol and centrifuged again (5 min, $21,000 \times g$, 4 °C).

6. The DNA pellets are dried for 10 min at room temperature and resuspended in 100 µL of buffer TE.

7. 10 µL of each DNA sample (excluding sample F) is linearized by digesting with NcoI overnight at 37 °C in a total volume of 30 µL.

8. Digested DNA samples (30 µL) are loaded onto a 1 % agarose gel and electrophoresed for 3 h at 130 V. Southern blot analysis is performed to analyze the enrichment of the DNA fragment of interest throughout the purification (Fig. 1b, *see* **Note 14**) [10, 13, 24].

3.4.2 Protein Analysis

1. Add 10 µL of 3× SDS sample buffer to each 20 µL protein sample.

2. Separate proteins by electrophoresis in a 10 % SDS-PAGE gel and transfer them to a membrane.

3. Perform a Western blot to detect the LexA-TAP fusion protein using goat anti-CBP primary antibody (1:200) followed by donkey anti-goat HRP secondary antibody (1:5,000) [10, 13, 24].

3.5 Preparation of DNA for Electron Microscopy

3.5.1 Preparation of Ethidium Bromide Carbon (EBC) Grids

1. 200 mL of distilled water is added to a Mechanical Ultrasonic Cleaner water bath placed in a fume hood.

2. A vial of 400-mesh copper electron microscopy grids is emptied into a clean 250 mL glass beaker.

3. 100 mL of acetone is added to the beaker, submerging the copper grids. The beaker is then placed in the Mechanical Ultrasonic Cleaner water bath and sonicated for 10 s in a fume hood.

4. The acetone is poured off and 100 mL of 100 % ethanol is added. The beaker is placed in the Mechanical Ultrasonic Cleaner water bath and sonicated for 10 s in a fume hood.

5. The ethanol is poured off and the glass beaker is inverted on a piece of Whatman filter paper and allowed to dry overnight. The grids will stick to the side of the beaker and drop onto the filter paper as the ethanol evaporates.

6. A carbon evaporator is prepared by thorough cleaning. Freshly sharpened carbon rods should be used for evaporation.

7. Freshly cleaved 1.5 cm squares of mica are placed on Whatman filter paper on the center support of the evaporator, 12 cm directly underneath the carbon rod source (*see* **Note 15**).

8. The chamber is sealed and the vacuum is pumped down overnight to $<1 \times 10^{-5}$ Torr.

9. The voltage is increased until carbon deposition measures ~7–10 nm, avoiding sparking from the carbon source.

10. The evaporator is turned off and the carbon-coated mica is stored in a desiccator until needed.

11. A Smith Grid Coating Trough is filled with Milli-Q water.

12. A rectangular piece of Whatman filter paper (1.5 cm × 3 cm) is cut and placed on top of the mesh support in the water bath. Additional Milli-Q water is added to fully submerge the filter paper.

13. Cleaned copper grids are placed face-up on the paper using fine-tipped tweezers (*see* **Note 16**). Each piece of carbon-coated mica can be used to generate approximately 20 carbon-coated grids if they are placed closely on the filter paper.

14. A piece of carbon-coated mica is picked up by the edge and gently lowered into the Grid Coating Trough at a 45° angle. The carbon layer will detach from the mica and float on top of the water.

15. Once the carbon film is completely free from its mica support, the positions of the filter paper and the carbon film are adjusted with tweezers to maximize the coverage of the carbon on the grids.

16. The water is slowly removed from the Grid Coating Trough with the syringe attachment so that the carbon film settles down on top of the grids.

17. The filter paper containing the grids is gently lifted out of the water bath, placed in a 10 cm petri dish lined with dry Whatman filter paper, and the dish is covered and placed in a desiccator overnight.

18. A fresh working stock of 30 μg/mL ethidium bromide is made in deionized water and is used directly without filtration.

19. 20 μL drops of ethidium bromide (30 μg/mL) are pipetted onto a fresh sheet of Parafilm.

20. Dried carbon-coated grids are floated on the drops, carbon side down, for 15 min.

21. Excess ethidium bromide is wicked away with filter paper and the grids are dried, carbon side up, for 10 min.

22. The ethidium bromide carbon grids should be used within 2 h.

3.5.2 Denaturing DNA Samples

1. 1–2 µL of cross-linked DNA (*see* Subheading 3.3, **step 23**) is brought to 3 µL with buffer TEN. 9.3 µL of deionized formamide is added to the DNA (70 % final) followed by 1 µL of glyoxal (0.66 M final).

2. The mixture is pipetted gently a few times (do not vortex) before incubating for 30 min in a 37 °C water bath, and then placed immediately on ice.

3. 5 µL of TEN buffer is added to each sample for a total volume of 18.3 µL (which dilutes the formamide to 50 %). The sample is kept on ice until spreading.

3.5.3 Spreading Denatured DNA Samples

1. A working stock solution (20 µL) of 0.02 % BAC is made in TEN buffer.

2. 1 mL of fresh working stock solution of 0.5 mM uranyl acetate is made in 95 % ethanol.

3. For every sample, a 10 cm petri dish is fitted with a 3 cm × 2 cm mica ramp taped to the rim such that the ramp is at a 45° angle into the dish. Avoid direct contact of the tape with the water in the dish, as it will disrupt the spreading (Fig. 2a).

4. 25 mL of freshly filtered Milli-Q water is added to each petri dish.

5. One small dish (for instance, the cap of a 15 mL conical tube) is filled with 1 mL of 0.5 mM uranyl acetate solution. Two additional 5 cm petri dishes are also set up, each containing 5 mL of 95 % ethanol for post-staining washes.

6. The water surface in the 10 cm dish is lightly dusted with graphite powder (Fig. 2a).

7. While waiting for the graphite to stop moving, 1 µL of the 0.02 % BAC spreading solution is added to the 18.3 µL denatured DNA sample on ice.

8. The solution is gently mixed by pipetting a few times; do not vortex.

9. 5 µL of the DNA/BAC solution are applied to the center of the mica ramp (Fig. 2a).

10. The DNA/BAC droplet is allowed to roll down the mica ramp and spread out on the surface of the water, forming a hypophase that will push back the graphite powder (Fig. 2b). The hypophase film will be several centimeters in diameter and is invisible except for the boundaries that are defined by the displaced graphite powder.

11. The hypophase is allowed to stabilize for 1 min before proceeding.

12. EBC grids are briefly touched, carbon side down, to the surface of the hypophase. The DNA may be picked up anywhere

Fig. 2 DNA spreading and shadowing techniques for electron microscopy. (**a**) A petri dish is filled with Milli-Q water, and a 2 cm × 3 cm mica ramp is secured to the side of the dish at a 45° angle with tape attached to the back. Graphite powder is dusted on the surface and denatured DNA is pipetted onto the middle of the ramp. (**b**) The graphite powder is pushed back by the hypophase formed on the surface of the water. After 1 min a grid is touched to the surface of the hypophase in the area outlined with a *dashed line*, which contains the highest density of DNA. The grid is processed further (Subheading 3.5.3, **step 13**). (**c**) Rotary metal shadowing (Subheading 3.5.4). The front view of the setup illustrates the location of the grids (*white dots*) on a glass slide which is attached to the rotary platform. Optimal shadowing is achieved on grids closest to the center of the slide. (**d**) A side view of the evaporator illustrates the optimal shadowing angle relative to the tungsten basket containing the platinum–palladium wire. Note that the rotary platform is essentially perpendicular with the base of the evaporator, which results in a 3–7° angle between the grids and the tungsten basket. Larger angles lead to excess platinum deposition on the grids, which reduces the contrast of the DNA in the EM

throughout the hypophase, but is more concentrated along the graphite boundary farthest away from the mica ramp (Fig. 2b).

13. Without wicking away any liquid, the grid is submerged, carbon side up, in 0.5 mM uranyl acetate for 20 s.

14. The grid is washed briefly twice by submerging it successively in two petri dishes containing 95 % ethanol.

15. Excess ethanol is wicked away and the grid is dried, carbon side up, on filter paper for 10 min (*see* **Note 17**).

16. The grids are stored in a grid box prior to rotary metal shadowing.

3.5.4 Rotary Metal Shadowing

1. The evaporator is prepared by replacing the components for carbon evaporation with those for metal evaporation, including addition of the rotary stage (*see* **Note 18**).

2. A strip of double-sided tape is placed on a standard microscope slide. Grids containing DNA are gently placed on the slide such that there is a small overlap (1/4 of the grid) with the double-sided tape. Do not place more than 1/4 of the grid on the tape, or it will rip when removed.

3. A small piece (1 cm²) of double-sided tape is placed on the bottom of the slide, which is then stuck to the center of the rotating platform (Fig. 2c).

4. The angle of the rotating platform in the evaporator is adjusted such that it is ~4° to the tungsten source filament (Fig. 2d) (*see* **Note 19**).

5. A new tungsten basket filament is placed in the evaporator. A new basket must be used every time because molten platinum weakens the tungsten, which will break if used more than once.

6. A 7 cm piece of 0.2 mm platinum–palladium (80:20) wire is cut. The wire is balled up by twirling it around the tip of a pair of fine-tip tweezers, and is then placed in the tungsten basket.

7. All the rubber seals of the evaporator are wiped with water and dried with lint-free Kimwipes. This step improves the vacuum seal dramatically. The chamber is then pumped for 4 h or to $\sim 1 \times 10^{-5}$ Torr.

8. Once the proper vacuum is reached, the rotating platform is turned on at 60 rpm.

9. The current box, which controls the current through the tungsten filament, and the metal depth-measuring device are turned on.

10. The evaporation shield is put in place to prevent the initial burst of evaporated platinum from hitting the grids. This causes significant background issues and decreases the contrast of the DNA.

11. Wearing welding goggles, the current is slowly increased to 18 mA, pausing for 30 s to allow the vacuum to recover.

12. Once the vacuum recovers, the current is increased slowly by 1 mA every 5 s until reaching 25 mA. The platinum should melt around 25–28 mA.

13. The vacuum is allowed to recover for 10 s following platinum melting. The evaporation shield is removed so that newly evaporated platinum will begin to coat the grids.

14. The current is increased by 1 mA every 5 s until reaching 35 mA. The vacuum is allowed to recover for 10 s before continuing.

15. The current is then increased slowly (1 mA every 5 s) to 40 mA. The platinum will begin to evaporate more rapidly in the 36–40 mA range and should ideally deposit 1 nm of platinum every second. This rate of deposition is maintained by slowly increasing the current when the rate starts to drop. The vacuum will be irreversibly weakening and it is critical to get as much metal deposition as possible before the vacuum cuts out.

16. Ideal platinum evaporation, as measured by the metal depth-measuring device, is ~100–120 nm thick. This range yields good contrast for single-stranded DNA.

17. The metal-shadowed grids are stored in a grid box until ready for electron microscopy.

18. The evaporator is cleaned with Milli-Q water and lint-free Kimwipes.

3.6 Electron Microscopy

1. Grids are examined in a JEOL 120 keV or similar TEM.

2. Images are taken at 20,000× magnification, with the "nm" scale incorporated into the raw .dm3 image files.

3. Images of control DNA molecules should be taken as well, preferably in fields that also contain sample molecules to provide size standards and a visual readout of psoralen cross-linking efficiency.

4. Images of more than 200 individual gene molecules should be captured and saved to allow the generation of accurate probability density distributions for measured states.

3.7 Single Gene Molecule Analysis

3.7.1 Tracing Single Gene Molecules

1. Raw image files (.dm3) are opened in ImageJ.

2. ImageJ is mirrored from an Apple laptop to an iPad with the "Air Display Host" program installed on the laptop and the "Air Display" App installed on the iPad.

3. Molecules are traced in ImageJ with a fine-tipped stylus using the "Freehand Line" option (Fig. 3a). Avoid moving the stylus too quickly, as this reduces the number of individual points along the trace.

4. Tracing must begin on one single-stranded arm of the V-end and terminate on the V-end of the other arm. The trace alternates between single-strand bubble sides throughout the molecule to generate crossovers at the beginning and end of every bubble. Crossing over increases the accuracy of the

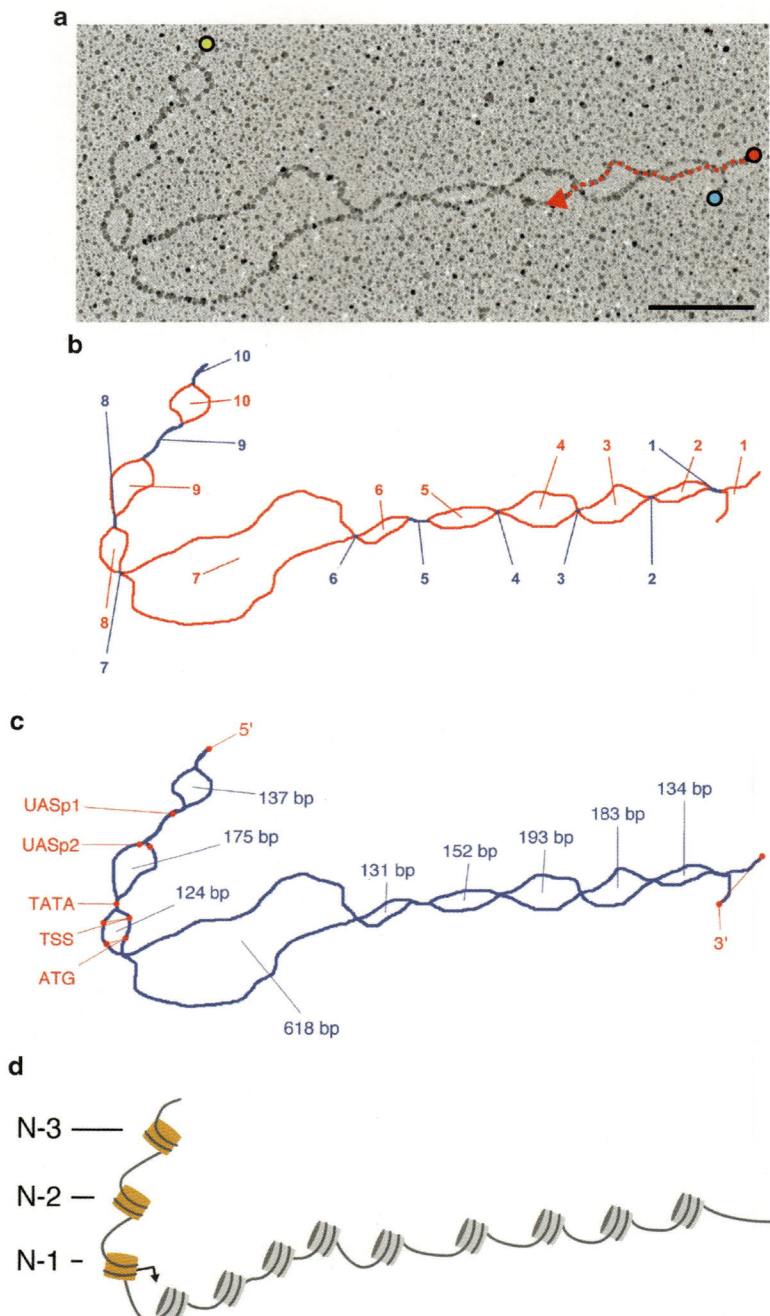

Fig. 3 Single gene molecule analysis with MATLAB. (**a**) Electron micrographs are opened in ImageJ and traced with a stylus. Traces begin on one arm of the 3′ "V-end" (*red circle with black outline*) and alternate between single-stranded sides of DNA bubbles (*dotted red line with red arrow*). The trace is continued to the 5′ end of the molecule (*yellow circle with black outline*) and doubles back along the molecule on the untraced sides of the DNA bubbles. The trace is terminated

bubble assignments in the custom MATLAB analysis program (*see* Subheading 3.7.2).

5. The length of the molecule (in nm) is measured, and the trace is exported as an X-Y coordinate .TXT file.

3.7.2 MATLAB Analysis of Single Gene Molecules

The X-Y coordinate trace data is imported into a custom program written in MATLAB code, where its features such as nucleosome occupancy at specific regulatory sites and R-values for different regions are analyzed and automatically calculated. R-values are the ratio of single-stranded DNA length to total DNA length, i.e., the fraction of base pairs that were nucleosome-associated. This automated analysis has the benefit of greatly increasing the throughput, minimizing the bias and measurement error associated with manual analysis, and additionally allowing for very rapid reanalysis of the same molecules if new questions are raised subsequently. Further, the software allows manual definition of properties of molecules (length, promoter location, etc.) and other parameters by the user, and thus may be useful in subsequent studies of other molecules (Fig. 3).

Briefly, the software analyzes a data trace by first determining which trace points correspond to single-stranded DNA (nucleosome sites) or double-stranded DNA (linker regions). This is automatically achieved by computing for all pairs of trace coordinates: (1) their distance along the length of the trace and (2) their direct Euclidean distance in the X-Y plane. Any points that lie beyond a minimum threshold Euclidean distance from all other points, except from nearby points along the trace, are assumed to fall in single-stranded DNA regions of the trace. All remaining points then are categorized into double-stranded DNA regions (Fig. 3b). The program then allows for user input to correct any errors in the automatic single-stranded DNA detection algorithm. Once the trace data is thus categorized, the software generates a lookup table defining whether each base pair in the DNA molecule is in a single-stranded or double-stranded DNA section. This lookup table then allows one to query the nucleosome occupancy at any position of interest (e.g., TATA box, Transcription Start Site (TSS), or ATG) or to compute the R-values for any region of interest (e.g., the ORF or promoter) (Fig. 3c). Knowledge of bubble locations and

Fig. 3 (continued) on the other single-stranded arm of the 3′ "V-end" (*blue circle with black outline*). Scale bar 100 nm. (**b**) The X-Y coordinate trace file is opened in MATLAB and analyzed. Single-strand DNA bubbles (*red*) and double-strand linker DNA regions (*blue*) are identified and numbered. This image is an example of the "numbered.png" file created for each molecule analyzed by the program. (**c**) The molecule is analyzed further by assigning bubble sizes (*blue*) and DNA features (*red*) (here, the 5′ end, UASp1, UASp2, TATA, TSS, ATG, and 3′ end). Other DNA features can be added depending on the molecule being analyzed. This image is an example of the "final.png" file created for each molecule analyzed. (**d**) A cartoon (created in Adobe Illustrator) of the nucleosome structure of the *PHO5* gene molecule in (**a–c**). The promoter is fully occupied with nucleosomes (N-1, N-2, and N-3, orange nucleosomes)

sizes allow the user to recreate cartoons of the original nucleosome structure in other software platforms such as Adobe Photoshop or Adobe Illustrator (Fig. 3d).

The program requires definition of several parameters for analysis of single gene molecules. These include the total length in base pairs, the base pair coordinates of DNA features (specific regulatory elements), and the regions for which R-values will be calculated. Additional parameters used in automatic bubble detection include the minimal number of points needed to be called a bubble, the maximum distance between two bubble endpoints to be considered part of the same bubble, the minimum Euclidean distance between two DNA strands to be considered in a bubble, and the minimum distance between two points along the X-Y trace to be considered in a bubble.

1. By default, the above parameters are set for *PHO5* gene molecules but they can be adjusted to analyze other molecules. The parameters can be adjusted directly in the software, or can be externally defined in several .TXT files. External definition is useful when applying the same parameters to an entire batch of trace files, and is necessary if using the batch reanalysis functionality (*see* **step 8** below). If any of the following files are located in the same folder as the X-Y coordinate file, the parameters will be loaded and override the software defaults:

    (a) "detection_parameters.txt", which includes the total base pair length of the molecule and parameter values for automatic bubble detection (used in **step 3** below).

    (b) "site_definitions.txt", which contains a list of DNA features with base pair coordinates. Nucleosome occupancy will be checked at these sites.

    (c) "region_definition.txt", which includes base pair windows that correspond to specific regions of the molecule, such as the promoter or ORF. R-values will be calculated for these regions.

2. An X-Y coordinate file is opened in the program by selecting "Load New Trace File" from the user interface window. The trace points are plotted.

3. After selecting "Detect Bubbles" in the user interface window, the program makes an attempt to automatically distinguish single-stranded DNA bubbles from double-stranded DNA linker regions.

4. A plot of the automatic bubble detection result is shown (Fig. 3b). Bubbles will appear red while linker DNA regions will appear blue. Bubbles and linker DNA will be labeled in numerical order starting from the 3' end of the molecule, identifiable by the single-stranded DNA "V-end."

5. Bubble detection parameters can be manually adjusted such that the program correctly identifies bubbles on the single gene molecule.

6. If needed, errors may be manually corrected by choosing "Correct Errors" from the user interface window. This is necessary, for example, if a bubble is erroneously indicated as a linker region, or if two ID numbers are assigned to a single bubble.

7. Following error correction, the "Analysis" option is selected and the molecule is processed (Fig. 3c). The program saves several new files in a new folder for each molecule analyzed. These include:

   (a) "analysis_info.txt", which reports the date, time, and software version number.

   (b) "bubble_bp_sizes.txt", which reports the base pair size of the bubbles, starting at the 5′ end of the molecule and ending with the 3′ "V-end".

   (c) "final.png", an image file showing the DNA features mapped to the X-Y trace data (Fig. 3c).

   (d) "molecule_info.txt", the lookup table with the X-Y trace data point assigned to each base pair in the molecule, and the bubble or linker assignment for each base pair.

   (e) "numbered.png", an image file showing the numbered bubbles and linkers to identify specific regions (Fig. 3b).

   (f) "results.txt", which reports if any of the DNA features are in a bubble, and if so, the base pair size of that bubble.

   (g) "trace_bubble_info.txt", which aids in reprocessing old trace data.

   (h) "trace_data.txt", a copy of the original X-Y trace coordinates.

8. Previously analyzed molecules may be revisited and batch processed using the "Batch Process Old Traces" option in the user interface window. Batch processing uses previously determined bubble and linker DNA assignments, avoiding the need to repeat bubble detection or error correction, while simultaneously changing DNA feature calls and R-value domains for different analysis questions that may arise.

4   Notes

1. Strains of *S. cerevisiae* and plasmid pSH17 (K2049) are available from the authors.

2. Strains containing plasmid pSH17 are grown in glucose-containing media when propagating the cells or making precultures, because glucose is a potent repressor of the *GAL1-10*

promoter and inhibits expression of the R recombinase gene. Raffinose is used as the carbon source when growing larger cultures to purify gene rings, because it does not interfere with the galactose-mediated induction of the R recombinase.

3. Cells growing under these conditions have a doubling time of ~2.5 h.

4. 20 mL syringes are prepared by removing the plunger and sealing the plastic tip by melting it closed with a Beckman Tube Topper. Alternatively, syringes with Luer fittings may be used and sealed with reusable female Luer plugs (Bio-Rad).

5. Rigorous shaking of the grinder prevents the sample from sticking to its sides. Cryo gloves should be worn to protect the hands during grinding.

6. If the dry ice has not evaporated enough the Extraction Buffer will partially freeze, creating a thick slurry. This should be avoided; however, if it happens, wait until the slurry returns to a mostly liquid state and start the 30 min timer at that point. Depending on the amount of yeast spaghetti, the incubation time for dry ice evaporation in **step 7** can be increased to avoid this issue.

7. Rings of different sizes require different centrifugation times. For example, *PHO5* gene rings (2.2 kb) are spun for 2.5 h, smaller rings (*PHO5* promoter rings (0.7 kb) and yeast 5S rDNA rings (0.75 kb)) for 3 h, and larger rings (full length yeast rDNA (9 kb)) for 2 h.

8. PolyPrep columns may leak from the caps during long incubations on rotary wheels, and should be fortified by placing a thin strip of Parafilm around the threads of the cap prior to placing them on the columns. A fresh strip of Parafilm should be used every time the cap is removed.

9. While collecting this sample, a 1 mL pipette tip with a larger opening is recommended over a smaller 200 µL pipette tip so that the resin/beads are collected efficiently for downstream Southern blot evaluation.

10. It is important to include an internal control for both psoralen cross-linking efficiency and for size comparisons when analyzing the images. Ideally, no single-stranded DNA bubbles should be seen on control molecules, which would be indicative of highly efficient psoralen cross-linking. The control molecules should also be of a different base-pair length than the sample to be analyzed so that they are easily distinguishable in the images. For example, in chromatin preparations of the *PHO5* gene (2.2 kb) we include a *PHO5* promoter fragment (0.65 kb) and a *PHO5* ORF fragment (1.6 kb) which are easily

distinguishable due to their size difference and their highly cross-linked appearance.

11. Psoralen cross-links double-stranded DNA when exposed to 366 nm UV light. However, the cross-links can be broken by 254 nm UV light, the standard wavelength provided in most cross-linking equipment (such as the Stratalinker and common handheld UV lamps for analysis of ethidium bromide-stained DNA in agarose gels). The ability to break psoralen cross-links is an essential part of being able to probe DNA on Southern blots of psoralen cross-linked samples because radiolabeled DNA probes cannot bind to psoralen-cross-linked DNA, resulting in weak or no signal.

12. Washing the petri dish with IRN buffer increases sample yield and reduces the ethanol concentration of the sample by 50 % such that downstream RNAse A and proteinase K steps are not adversely affected.

13. Circularized gene rings are supercoiled and will be tightly packed on electron microscopy grids. In order to visualize single-stranded DNA bubbles they must first be relaxed by either nicking or cutting with restriction enzymes. Nicking a single strand will relax the ring and reveal the average nucleosome density but will not contain any orientational information (for example, the location of the promoter versus the ORF) because the DNA is still circular. NcoI is used for *PHO5* gene rings because it cuts in a single location between the LexA binding cluster and the 3′ RS element. The linearized molecules can then be oriented because the LexA binding cluster is protected from cross-linking, resulting in a 3′ single-stranded DNA "V-end" downstream of the ORF [9]. The LexA "V-end" has also been observed in DNA from purified rDNA chromatin rings after psoralen cross-linking, DNA isolation, and NcoI digestion [24].

14. Ethidium bromide must not be added to gels for psoralen-cross-linked samples, as it will skew the migration of the bands. Prior to transferring the sample to a nitrocellulose membrane it is important to UV irradiate (254 nm) the gel on both sides for 48 s each in a Stratalinker 2400 to break interstrand psoralen cross-links, allowing radiolabeled DNA probes to hybridize to their complementary sequences.

15. Place a coin on the filter paper such that at least one edge of each piece of mica is covered. This prevents movement of the mica due to vacuum fluctuations following carbon deposition.

16. The "face-up" side of copper grids is completely shiny while the "face-down" side is matted and only shiny on the outer rim.

17. Tweezers should be wiped dry between handling EBC grids, or the residual ethanol will disrupt the hypophase when picking up the next sample. Up to five grids may be used for each spread, and each sample can be spread three times (5 µL for each spread). Subsequent spreads should be performed using new petri dishes, mica ramps, and graphite dusting. The BAC solution should only be added once to each DNA sample, and not for each subsequent spread. Denatured DNA samples containing BAC are not stable for long-term storage and should be discarded if not used within an hour.

18. The vacuum should be cycled once or twice (overnight pumping is best) before shadowing. This conditions the equipment and ensures a good vacuum during shadowing.

19. The tungsten filament in the evaporator utilized by our lab is 1 cm off-center and 13 cm above the center of the rotating platform (Fig. 2c–d). When the surface of the platform is perpendicular to the benchtop (facing towards the user), the angle between the surface and the filament is 4.4° (Fig. 2c–d). The optimum angle for rotary shadowing is between 3 and 7° [23].

Acknowledgements

This work was supported by an NIH NRSA grant to C.R.B. (F32GM087867) and an NSF grant (1243957) to H.B.

References

1. Mao C, Brown CR, Griesenbeck J et al (2011) Occlusion of regulatory sequences by promoter nucleosomes in vivo. PLoS One 6:e17521

2. Almer A, Rudolph H, Hinnen A et al (1986) Removal of positioned nucleosomes from the yeast PHO5 promoter upon PHO5 induction releases additional upstream activating DNA elements. EMBO J 5:2689–2696

3. Kornberg RD (1974) Chromatin structure: a repeating unit of histones and DNA. Science 184:868–871

4. Macaulay IC, Voet T (2014) Single cell genomics: advances and future perspectives. PLoS Genet 10:e1004126

5. Raser JM, O'Shea EK (2004) Control of stochasticity in eukaryotic gene expression. Science 304:1811–1814

6. Boeger H, Griesenbeck J, Kornberg RD (2008) Nucleosome retention and the stochastic nature of promoter chromatin remodeling for transcription. Cell 133:716–726

7. Jessen WJ, Hoose SA, Kilgore JA et al (2006) Active PHO5 chromatin encompasses variable numbers of nucleosomes at individual promoters. Nat Struct Mol Biol 13:256–263

8. Mao C, Brown CR, Falkovskaia E et al (2010) Quantitative analysis of the transcription control mechanism. Mol Syst Biol 6:431

9. Brown CR, Mao C, Falkovskaia E et al (2013) Linking stochastic fluctuations in chromatin structure and gene expression. PLoS Biol 11:e1001621

10. Griesenbeck J, Boeger H, Strattan JS et al (2003) Affinity purification of specific chromatin segments from chromosomal loci in yeast. Mol Cell Biol 23:9275–9282

11. Griesenbeck J, Boeger H, Strattan JS et al (2004) Purification of defined chromosomal domains. Meth Enzymol 375:170–178

12. Ansari A, Cheng TH, Gartenberg MR (1999) Isolation of selected chromatin fragments from yeast by site-specific recombination in vivo. Methods (San Diego, CA) 17:104–111

13. Hamperl S, Brown CR, Perez-Fernandez J et al (2014) Purification of specific chromatin domains from single-copy gene loci in Saccharomyces cerevisiae. Methods Mol Biol 1094:329–341

14. Hanson CV, Shen CK, Hearst JE (1976) Cross-linking of DNA in situ as a probe for chromatin structure. Science 193:62–64

15. Cech TR, Pardue ML (1976) Electron microscopy of DNA crosslinked with trimethylpsoralen: test of the secondary structure of eukaryotic inverted repeat sequences. Proc Natl Acad Sci U S A 73:2644–2648

16. Cech TR, Pardue ML (1977) Cross-linking of DNA with trimethylpsoralen is a probe for chromatin structure. Cell 11:631–640

17. Cech TR, Potter DA, Pardue ML (1977) Electron microscopy of DNA cross-linked with trimethylpsoralen: a probe for chromatin structure. Biochemistry 16:5313–5321

18. Musajo L, Rodighiero G (1970) Studies on the photo-C4-cyclo-addition reactions between skin-photosensitizing furocoumarins and nucleic acids. Photochem Photobiol 11:27–35

19. Cole RS (1970) Light-induced cross-linking of DNA in the presence of a furocoumarin (psoralen). Studies with phage lambda, Escherichia coli, and mouse leukemia cells. Biochim Biophys Acta 217:30–39

20. Toussaint M, Levasseur G, Tremblay M et al (2005) Psoralen photocrosslinking, a tool to study the chromatin structure of RNA polymerase I-transcribed ribosomal genes. Biochem Cell Biol 83:449–459

21. Sogo JM, Stahl H, Koller T et al (1986) Structure of replicating simian virus 40 minichromosomes. The replication fork, core histone segregation and terminal structures. J Mol Biol 189:189–204

22. de Bernardin W, Koller T, Sogo JM (1986) Structure of in-vivo transcribing chromatin as studied in simian virus 40 minichromosomes. J Mol Biol 191:469–482

23. Sogo JM, Thoma F (1989) Electron microscopy of chromatin. Meth Enzymol 170:142–165

24. Hamperl S, Brown CR, Garea AV et al (2014) Compositional and structural analysis of selected chromosomal domains from Saccharomyces cerevisiae. Nucleic Acids Res 42:e2

25. Lucast LJ, Batey RT, Doudna JA (2001) Large-scale purification of a stable form of recombinant tobacco etch virus protease. Biotechniques 30:544–546

Chapter 10

Purification of Proteins on Newly Synthesized DNA Using iPOND

Huzefa Dungrawala and David Cortez

Abstract

The replisome is a large protein machine containing multiple enzymatic activities needed to complete DNA replication. In addition to helicase and polymerases needed for copying the DNA, the replisome also contains proteins like DNA methyltransferases, histone chaperones, and chromatin modifying enzymes to couple DNA replication with chromatin deposition and establishment of the epigenetic code. In addition, since template DNA strands often contain DNA damage or other roadblocks to the replication machinery, replication stress response proteins associate with the replisome to stabilize, repair, and restart stalled replication forks. Hundreds of proteins are needed to accomplish these tasks. Identifying these proteins, monitoring their posttranslational modifications, and understanding how their activities are coordinated is essential to understand how the genome and epigenome are duplicated rapidly, completely, and accurately every cell division cycle. Here we describe an updated iPOND (isolation of proteins on nascent DNA) method to facilitate these analyses.

Key words DNA replication, Chromatin, DNA repair, iPOND, EdU, Click chemistry, Replication stress, PCNA, DNA damage

1 Introduction

Studying proteins at active or damaged replication forks and monitoring the deposition of proteins on newly synthesized DNA requires a method of purifying these protein–DNA complexes. Chromatin fractionation and chromatin-immunoprecipitation (ChIP) provide two methods to accomplish this task. However, chromatin fractionation does not provide any spatial information about the location of where a protein is bound in the genome. ChIP provides spatial information but its utility in studying DNA replication or replication stress responses in mammalian cells is limited by difficulties in synchronizing replication in any specific genomic region. Super-resolution immunofluorescence imaging or analysis of protein–protein interactions using methods like proximity ligation assays can provide spatial information with respect to a

Ronald Hancock (ed.), *The Nucleus*, Methods in Molecular Biology, vol. 1228,
DOI 10.1007/978-1-4939-1680-1_10, © Springer Science+Business Media New York 2015

known reference control. However, these methods require highly specific antibodies and are not compatible with unbiased approaches to identify new proteins or protein modifications.

We developed iPOND to overcome these experimental challenges, providing a method to examine protein recruitment and modification at replication forks as well as the processes of chromatin deposition and maturation [1, 2]. iPOND is essentially a reverse chromatin immunoprecipitation in which newly replicated DNA is purified and its associated proteins analyzed by immunoblotting or mass spectrometry [3, 4]. iPOND relies on incorporation of the nucleoside analog 5-ethynyl-2′-deoxyuridine (EdU) into newly synthesized DNA. EdU is incorporated instead of thymidine and contains an alkyne functional group permitting a high-efficiency cycloaddition reaction to tether biotin to the newly synthesized DNA fragment [5]. Biotin facilitates a single-step, streptavidin-based affinity purification of the DNA–protein complexes, which can then be analyzed using standard DNA and protein detection methods.

The procedure is most useful when cells are labeled with EdU for very short periods (2–15 min) so that it is incorporated into only small DNA fragments immediately adjacent to elongating replication forks. Fixation of protein–DNA complexes using formaldehyde prior to performing the biotin-conjugation reaction and purification permits isolation of replisome components (Fig. 1a). Combining the short EdU labeling period with increasing culture times in the absence of label (a chase period), is useful to study chromatin deposition and maturation as a function of time and distance from the elongating replication fork (Fig. 1b). Finally, combining the EdU labeling with addition of drugs that damage DNA or otherwise stall replication forks such as hydroxyurea provides a method to study DNA repair and replication stress responses (Fig. 1c).

2   Materials

Culture mammalian cells in appropriate growth medium and conditions. The protocol described here is optimized for HEK293T cells grown in DMEM containing 7.5 % fetal bovine serum. However, other cell types including U2OS, hTERT-RPE, and mouse embryonic fibroblasts have been used successfully [6]. All reagents are prepared and used at room temperature unless otherwise specified.

2.1 Labeling Cells with EdU and Fixation

1. EdU: 10 mM 5-ethynyl-2′-deoxyuridine (Life Technologies) stock solution prepared in DMSO. Store at –20 °C protected from light for up to 1 year.

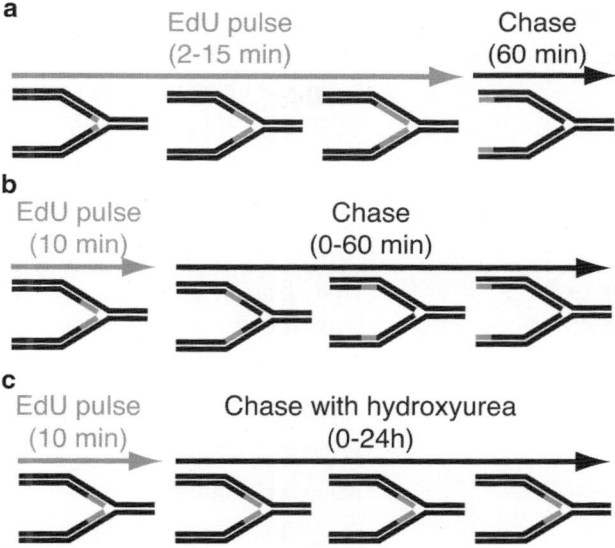

Fig. 1 Diagram of three types of iPOND experiment. In these schematics, the *black lines* represent unlabeled DNA and the *orange lines* represent DNA labeled with EdU. (**a**) Experiment to detect proteins that localize at elongating replication forks. Cells are incubated for increasing times in EdU prior to fixation. A single chase sample in which the EdU is removed for 1 h prior to fixation is included as a control. True replication proteins will no longer be enriched in this chase control. (**b**) Experiment to monitor chromatin deposition and maturation. A single short EdU labeling period is followed by increasing chase times to monitor how the proteins associated with the EdU-labeled DNA fragment change as a function of time and distance from the fork. (**c**) Experiment to assess replication stress responses. Cells are labeled with EdU and then chased in medium containing a replication stress-inducing drug like hydroxyurea. Variants of this procedure in which the replication stressing agent is added prior to EdU or in which EdU remains in the growth medium during the replication stress period are possible. All iPOND experiments should include an additional control sample that either lacks EdU labeling or in which the biotin azide is omitted from the procedure (a "no-click" control)

2. Thymidine: 10 mM thymidine (Sigma) stock solution prepared in water and stored at −20 °C.

3. Fixative and permeabilization buffer: immediately prior to use, dilute 37 % formaldehyde (Sigma) in phosphate buffered saline (PBS) solution (Thermo Fisher) to a final concentration of 1 %. Permeabilization buffer is prepared by diluting Triton X-100 (Sigma) to 0.25 % in PBS.

4. 1.25 M glycine (Sigma) in water.

5. Cell lifter (Sigma) or equivalent.

6. Cell wash solution: 0.5 % (w/v) bovine serum albumin (Sigma) dissolved in PBS.

2.2 Click Reaction Components	1. Biotin Azide (Life Technologies): 1 mM stock solution prepared in DMSO and stored at –20 °C.
	2. $CuSO_4$ (Thermo Fisher): 100 mM stock solution in water, store at room temperature.
	3. (+) sodium L-ascorbate (Sigma): prepare freshly a 20 mg/mL solution in water, limit exposure to air, and store on ice until needed.
2.3 Cell Lysis and Purification Components	1. Lysis buffer: 1 % SDS (Sigma) in 50 mM Tris-HCl, pH 8.0 with aprotinin and leupeptin (1 μg/mL each).
	2. 90 micron nylon mesh (Small Parts, Logansport, IN, USA).
	3. Streptavidin agarose (EMD Millipore).
	4. High salt wash: 1 M NaCl.
	5. Microtip sonicator.
	6. Streptavidin-agarose beads [sigma or equivalent).
2.4 Elution and Protein Detection Components	1. SDS sample buffer (2×): mix 0.4 g of SDS, 2 mL of 100 % glycerol, 1.25 mL of 1 M Tris, pH 6.8, and 0.01 g of bromophenol blue in 8 mL of water. Prior to use, add 1 M DTT solution to a final concentration of 200 mM.
	2. Cross-link reversal solution: mix 0.5 M EDTA, 1 M Tris–HCl, pH 6.7, proteinase K (500 units/mL) (Sigma) in a 2:4:1 volume ratio immediately prior to use.

3   Methods

3.1 Labeling of Cells	1. Expand HEK293T cells such that there are 4–6×10^7 cells per 150 mm-diameter culture dish. Three dishes are needed per sample. This protocol is written to assume three samples (*see* **Note 1**). Also prepare one extra dish of cells, so ten dishes are needed in total.
	2. Harvest the extra dish of cells and count the viable cells. This cell number is used to calculate the volume of reagents to use in subsequent steps.
	3. Remove the sample dishes from the incubator and add EdU to the culture medium to a final concentration of 10 μM for all three samples (*see* **Note 2**). Replace into the incubator rapidly and incubate for 10 min.
	4. For samples #1 and #2, remove the cells from the incubator and fix on the dish by adding 10 mL of 1 % formaldehyde solution. Incubate for 20 min at room temperature.
	5. For sample #3, remove the cells from the incubator after 10 min of labeling, decant the growth medium, carefully wash

the cells with 5 mL of growth medium containing 10 μM thymidine, decant, and replace with 20 mL of growth medium containing 10 μM thymidine before replacing the cells into the incubator for a further 60 min prior to fixing as in **step 4** (*see* **Note 3**).

6. Quench the cross-linking reaction by adding 1 mL of 1.25 M glycine solution.

7. Collect the cells by scraping with a cell lifter and transfer into a 50 mL conical tube. Place on ice. Combine the three plates per sample into a single tube. Record the volume.

8. Centrifuge 5 min at $900 \times g$, 4 °C.

9. Decant the supernatant.

10. Wash the cell pellets with the same volume of PBS as the fixation volume noted in **step 7**, vortexing to resuspend the pellets. Centrifuge for 5 min at $900 \times g$, 4 °C.

11. Decant the PBS wash and repeat the washing **step 10** two additional times.

12. After the final wash, decant the PBS (*see* **Note 4**).

13. Resuspend the cells in permeabilization buffer at a concentration of 1×10^7 cells/mL for 30 min at room temperature.

14. Centrifuge the cells for 5 min at $900 \times g$, 4 °C. Decant the supernatant.

15. Wash the cells once with 0.5 % BSA in PBS at 4 °C.

16. Centrifuge the cells for 5 min at $900 \times g$, 4 °C. Decant the supernatant.

17. Wash the cells once with PBS at 4 °C.

18. Centrifuge the cells 5 min at $900 \times g$, 4 °C. Decant the supernatant. Place the cells on ice.

3.2 Click Chemistry Reaction to Conjugate Biotin to EdU

1. Prepare the click reaction cocktail to contain a final concentration of 10 mM sodium ascorbate, 2 mM $CuSO_4$, and 10 μM biotin azide, in PBS. Prepare sufficient cocktail for 5 mL per 10^8 cells (*see* **Note 5**). Omit the biotin azide from the cocktail prepared for sample #1; this is the "no-click" negative control (*see* **Note 6**).

2. Resuspend the cell pellets in the appropriate click reaction cocktail at a final concentration of 2×10^7 cells/mL.

3. Rotate the samples at room temperature for 1–2 h.

4. Centrifuge the cells for 5 min at $900 \times g$, 4 °C. Decant the supernatant.

5. Wash the cells once with 0.5 % BSA in PBS at 4 °C using a volume equal to the click reaction cocktail volume used in **step 2**.

6. Centrifuge the cells for 5 min at $900 \times g$, 4 °C. Decant the supernatant.

7. Wash the cells once with PBS at 4 °C.

8. Centrifuge the cells for 5 min at $900 \times g$, 4 °C. Decant the supernatant (*see* **Note 7**).

3.3 Cell Lysis and Purification of Captured DNA–Protein Complexes

1. Resuspend the cells in lysis buffer containing aprotinin and leupeptin at a concentration of 1.5×10^7 cells per 100 μL. Transfer to 1.5 mL microfuge tubes and place on ice.

2. Sonicate using a microtip sonicator at a power of 13–16 W, 20 s pulse.

3. Pause 40 s with the samples on ice to prevent over-heating.

4. Repeat sonication **steps 2** and **3** four times (*see* **Note 8**).

5. Centrifuge the samples for 10 min at $16,100 \times g$, room temperature.

6. Filter the supernatant through a 90-μm nylon mesh into a new tube, place on ice (*see* **Note 9**).

7. Dilute the lysate 1:1 (v/v) using PBS containing 1 μg/mL each of aprotinin and leupeptin at 4 °C.

8. Save 15 μL of the lysate as an "input" sample.

9. Add 50 μL of packed streptavidin-agarose beads pre-washed with PBS (*see* **Note 10**).

10. Incubate for 1–16 h in the cold room on a rotating mixer (*see* **Note 11**).

11. Centrifuge the streptavidin-agarose beads containing the captured DNA–protein complexes for 3 min at $1,800 \times g$, room temperature.

12. Decant the supernatant.

13. Wash the beads with 1 mL of cold lysis buffer. Rotate for 5 min.

14. Centrifuge for 1 min at $1,800 \times g$. Decant the supernatant.

15. Wash with 1 mL of 1 M NaCl. Rotate for 5 min.

16. Centrifuge for 1 min at $1,800 \times g$. Decant the supernatant.

17. Repeat the washes with lysis buffer in **steps 15** and **16** two additional times.

18. Centrifuge for 1 min at $1,800 \times g$. Decant the supernatant.

3.4 Elution and Detection of Proteins Bound to Captured DNA

1. Add SDS sample buffer to the beads (1:1 v/v of packed beads). Also mix the "input" samples saved in **step 10** (Subheading 3.3) with an equal volume of SDS sample buffer.

2. Incubate all the samples for 25 min at 95 °C (*see* **Note 12**).

3. Centrifuge for 1 min at $1,800 \times g$, room temperature.

4. The purified protein samples and input samples are ready for separation by standard SDS-PAGE (*see* **Note 13**).

5. Transfer the proteins to nitrocellulose and detect by standard immunoblotting procedures (*see* **Notes 14** and **15**).

6. As an alternative to immunoblotting, proteins can be detected and analyzed by mass spectrometry (*see* **Note 16**).

4   Notes

1. A typical experiment to examine a protein at an elongating replication fork will require three samples (nine dishes). Sample #1 is used as a negative control for the purification. Sample #2 purifies proteins at or immediately adjacent to the replication fork. Sample #3 purifies general chromatin-binding proteins. This protocol is written for an experiment with these three samples.

2. Stagger the samples in time to ensure sufficient time for processing each sample throughout the experiment. EdU labeling times can vary and be as short as 2 min. Typically, we find that a 10 min labeling time is a useful starting point to purify replication proteins.

3. Pre-equilibrate the growth medium to 37 °C and the proper CO_2 content. Complete the wash step as rapidly as possible to minimize the length of time the cells are not in the incubator. The thymidine in this step is optional, but may decrease the chance of residual EdU being incorporated into DNA during the 1 h "chase" incubation.

4. We find it useful to flash-freeze cell pellets and store them at −80 °C when multiple samples are prepared over an extended staggered time.

5. A typical sample of three 150 mm dishes of HEK293T cells will have 1.5×10^8 cells and will need 7.5 mL of click reaction cocktail. In our growth conditions approximately 50 % of the cells in a culture are actively synthesizing DNA. We recommend increasing the cell number for cell types that have fewer cells in S-phase at the time of the experiment.

6. The "no-click" negative control can be replaced with a sample in which no EdU is used to label the cells.

7. Samples can be stored at −80 °C after this step.

8. The lysate should clarify somewhat during sonication. Cloudiness is an indicator of insufficient sonication or incorrect cell/lysis volume ratio. Larger volumes will require additional sonication steps.

9. The lysate should be clear after centrifugation and filtering. If not, repeat the centrifugation to remove any insoluble material.

10. Streptavidin-bound paramagnetic beads can be substituted. We have found that purification with magnetic beads using standard magnetic bead separation procedures can yield less nonspecific proteins in some cases. The volume of beads may need to be adjusted based on the biotin-binding capacity of beads from different manufacturers.

11. We have found that incubation for shorter times (1 h) typically yields good results. Extended incubation times may increase sensitivity.

12. This step reverses the protein–DNA cross-links and solubilizes the denatured proteins.

13. We usually use a 15 % SDS-PAGE gel to detect PCNA and histone proteins as controls. The purified sample can be split onto two or three gels to detect these positive controls easily by standard immunoblotting methods. Load all the sample onto a single gel for detecting proteins found more rarely at replication forks. We recommend loading the equivalent of 0.1 % (v/v) of the input to compare to the final purified samples (*see* Fig. 2).

14. It may be difficult to fully reverse the cross-links, especially for large proteins. Thus, you should expect that some proteins will be detected on SDS-PAGE gels at both their typical position and as higher molecular weight aggregates.

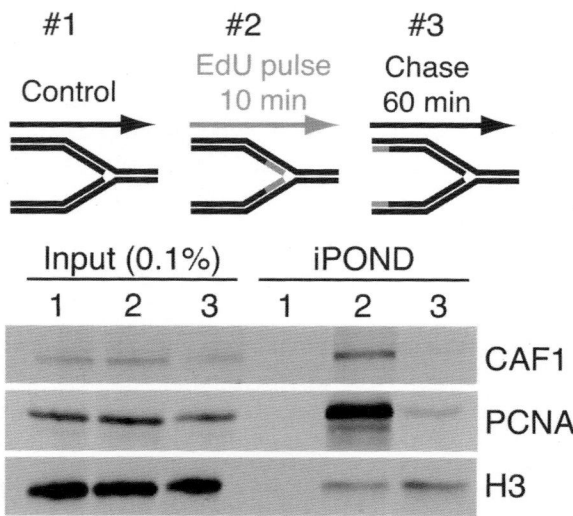

Fig. 2 An example of the results expected from a simple three-sample iPOND experiment (no-click control, 10 min EdU label, and 10 min EdU label followed by 1 h chase) (*upper panel*). *Lower panel*, immunoblotting to compare the abundance of CAF1 (chromatin assembly factor 1), PCNA (proliferating cell nuclear antigen) and histone H3 in the purified samples with that in the input cell lysate

15. *See* Fig. 2 for an example of the expected result from this simple, three sample experiment.

16. We typically use a "short stack" gel method prior to mass spectrometry. In this procedure, proteins are electrophoresed for a short time into the top 0.5 cm of the polyacrylamide gel. This gel piece is excised prior to standard in-gel trypsinization and mass spectrometry methods.

Acknowledgements

This work was supported by NCI grant R01CA136933 to D.C. We thank Bianca Sirbu and Frank Couch who contributed to the initial invention of iPOND.

References

1. Sirbu BM, Couch FB, Cortez D (2012) Monitoring the spatiotemporal dynamics of proteins at replication forks and in assembled chromatin using isolation of proteins on nascent DNA. Nat Protoc 7:594–605

2. Sirbu BM, Couch FB, Feigerle JT et al (2011) Analysis of protein dynamics at active, stalled, and collapsed replication forks. Genes Dev 25: 1320–1327

3. Lopez-Contreras AJ, Ruppen I, Nieto-Soler M et al (2013) A proteomic characterization of factors enriched at nascent DNA molecules. Cell Rep 3:1105–1116

4. Sirbu BM, McDonald WH, Dungrawala H et al (2013) Identification of proteins at active, stalled, and collapsed replication forks using isolation of proteins on nascent DNA (iPOND) coupled with mass spectrometry. J Biol Chem 288:31458–31467

5. Salic A, Mitchison TJ (2008) A chemical method for fast and sensitive detection of DNA synthesis in vivo. Proc Natl Acad Sci U S A 105:2415–2420

6. Nagarajan P, Ge Z, Sirbu B et al (2013) Histone acetyl transferase 1 is essential for mammalian development, genome stability, and the processing of newly synthesized histones H3 and H4. PLoS Genet 9:e1003518

Chapter 11

Applying the Ribopuromycylation Method to Detect Nuclear Translation

Alexandre David and Jonathan W. Yewdell

Abstract

Protein translation in the nucleus has been controversial for more than four decades. To take a new look at this potentially important phenomenon, we adapted the RiboPuromycylation Method (RPM) which labels actively translating ribosomes in cells via standard immunofluorescence microscopy. RPM is based on puromycylation of nascent chains trapped on ribosomes by antibiotics which inhibit chain elongation, followed by cell permeabilization/fixation and detection of puromycylated nascent chains using a puromycin-specific monoclonal antibody. To adapt the method to the nucleus, we use NP-40 rather than digitonin to permeabilize cells because NP-40 enables better antibody penetration into the nucleoplasm and particularly the nucleoli, a region of high translation as shown by RPM.

Key words Nuclear translation, Puromycin, Ribopuromycylation, Ribosome, Emetine, Nascent chain

1 Introduction

The translation of proteins in the eukaryotic nucleus (nuclear translation) was first reported 60 years ago [1, 2], but lost credence over the years as multiple lines of evidence pointed to translation being limited to the cytosol and mitochondria. Cook and colleagues in 2001 challenged the dogma by providing compelling evidence for protein synthesis in isolated nuclei [3]. Soon enough, these observations were largely attributed to contamination of nuclei with cytoplasmic ribosomes [4, 5], and the concept of nuclear translation was essentially reinterred. Given the complexity and mechanistic diversity of cell function, on general biological principles (where "never say never" is generally true and always useful in conceiving and interpreting experiments), nuclear translation should occur at some level in some eukaryotic cells. We became interested in nuclear translation as a mechanism to explain the curious features of the generation of peptides presented by major histocompatibility class I molecules [6].

Ronald Hancock (ed.), *The Nucleus*, Methods in Molecular Biology, vol. 1228,
DOI 10.1007/978-1-4939-1680-1_11, © Springer Science+Business Media New York 2015

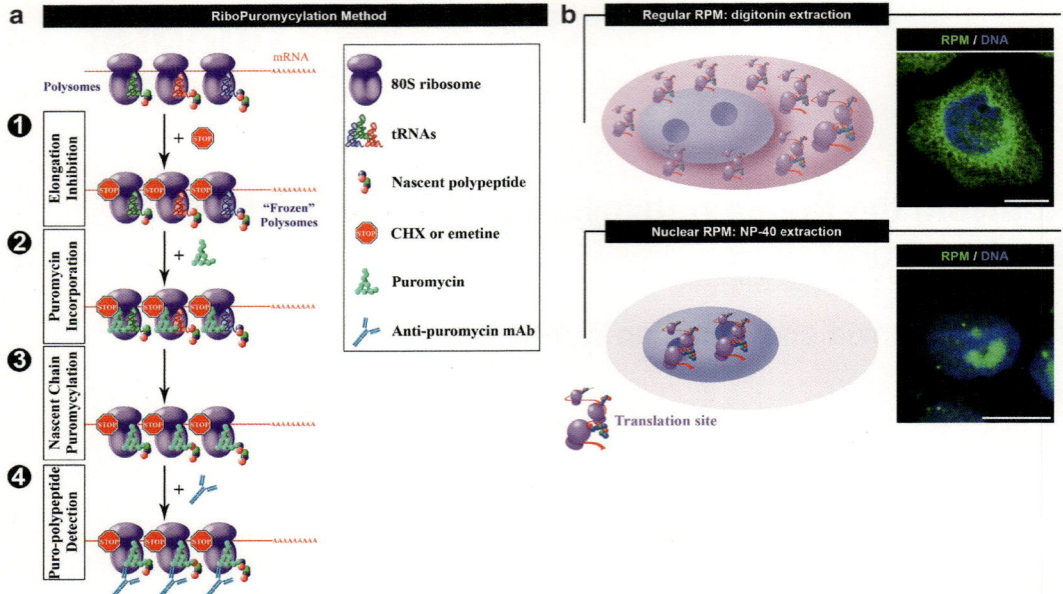

Fig. 1 (**a**) Schematic representation of RPM. After polysomes are frozen with an elongation inhibitor (*step 1*), puromycin is added (*step 2*) to living cells and nascent chains become puromycylated through ribosome catalysis (*step 3*). Anti-puromycin monoclonal antibodies detect the puromycylated nascent chains via indirect immunofluorescence (*step 4*). (**b**) Using the regular RPM protocol [7] we detect mostly cytoplasmic translation at the expense of nuclear translation, which represents only a minor fraction of global protein synthesis. NP-40 extraction washes away cytoplasmic polysomes and favors the detection of nuclear nascent chains. Images show living HeLa cells incubated with the translation inhibitor emetine together with puromycin and processed for either regular RPM (using digitonin) or nuclear RPM (using NP-40) prior to fixation. Scale bars = 10 μm. Reproduced from [7]

To extend these findings, we modified the RiboPuromycylation Method (RPM), which we developed to visualize nascent chains on ribosomes in permeabilized and fixed cells [7]. Puromycin (PMY) is an aminonucleoside antibiotic that mimics charged-tRNATyr to enter the ribosome A-site, where the ribosome catalyzes its covalent incorporation into the nascent chain's COOH-terminus [8]. As a result, translation is prematurely terminated and the ribosome dissociates, releasing the PMY-labeled nascent chain. By co-incubating with both puromycin and emetine, an irreversible inhibitor of chain elongation, we demonstrated that we were able not only to trap the puromycylated nascent chain within the ribosome but also to moderately improve puromycin incorporation in comparison with another elongation inhibitor, cycloheximide [9]. Labeling cells using a PMY-specific monoclonal antibody visualizes ribosomes with nascent chains, essentially caught in the act of translation (though stalled ribosomes, which feature prominently in neurons [10] and are likely to be present in other types of cells, are also detected) (Fig. 1a). Elongation inhibitors (e.g., cycloheximide and emetine) freeze nascent chains on ribosomes, which is the main advantage of RPM compared to traditional methods such as radiolabeling or incorporating modified amino acids where released proteins constitute an increasing bulk of the labeled protein with increasing time of incubation.

While developing the original RPM protocol, we frequently noted weak but specific labeling of the HeLa cell nucleoplasm. Antibody access to the nucleus (and particularly to the nucleolus) is limited when mild detergents like digitonin are used to permeabilize cells. To enable antibody access to the nucleus, we modified the original protocol by using NP-40 to extract cytoplasm prior to paraformaldehyde fixation (or to simultaneously permeabilize and fix) (Fig. 1b). The protocol described below greatly enhances the nuclear RPM signal, and with a number of critical controls provides robust evidence for nuclear translation. Nuclear RPM has been quickly employed by several groups to bolster independent lines of evidence that together support nuclear translation coupled to transcription in Drosophila [11] and MHC class I antigenic peptide generation from pre-spliced mRNAs [12].

2 Materials

2.1 Cells

HeLa cells (ATCC CCL-2.1). This procedure can be easily adapted for other cell lines as long as they possess adhesion capabilities similar to HeLa cells. Another procedure has been set up for non-adherent cells, though it does not remove cytoplasmic labeling [7].

2.2 Reagents

1. DEPC-treated water (Life Technologies).
2. EDTA-free protease inhibitor tablets (Roche).
3. NP-40 (Pierce).
4. Saponin (Sigma).
5. Paraformaldehyde (PFA): 16 % aqueous solution (Electron Microscopy Sciences).
6. Fluoromount-G (Southern Biotech).

2.3 Solutions

1. Dulbecco's Modified Eagle's Medium with glutamine, 100 µg/mL streptomycin, 100 U/mL penicillin, and 7.5 % fetal bovine serum (FBS).
2. Phosphate buffered saline (PBS): 210.0 mg/L KH_2PO_4, 9,000 mg/L NaCl, 726.0 mg/L $Na_2HPO_4 \cdot 7H_2O$.
3. Extraction buffer: 50 mM Tris–HCl, pH 7.5, 150 mM NaCl, 1 % (v/v) NP-40, EDTA-free protease inhibitor tablets (Roche).
4. Wash buffer: 50 mM Tris–HCl, pH 7.5, 150 mM NaCl, EDTA-free protease inhibitors (one mini tablet per 10 mL).
5. Labeling buffer: 0.05 % saponin, 10 mM glycine, 5 % FBS, 1× PBS, pH 7.5.

Table 1
Protein synthesis inhibitors

Inhibitor	Conditions	Reversible	Mechanism of action	Effect on ribosomes	Reference
Puromycin	50 µg/mL, 5 min	Yes	tRNA^{Tyr}-mimetic, enters A-site and associates covalently with nascent chain	Blocks elongation step of translation by inducing premature termination	[8]
Emetine	25 µg/mL, 5 min	No	Binds 40S ribosome subunit	Freezes translation during elongation; stabilizes polysomes	[13]
Anisomycin	10 µg/mL, 15 min	Yes	Binds A-site, 60S ribosome subunit. Competes with puromycin	Blocks peptidyl-transferase activity; stabilizes polysomes	[14]
Sodium arsenite	65 µg/mL, 15 min	Yes	Oxidative stressor; induces eIF2a phosphorylation	Induces stress granules; inhibits polysome formation	[15]
Harringtonine	2 µg/mL, 15 min	Partly	Prevents peptide bond formation at the initiation complex	Blocks initiation step of translation; inhibits polysome formation	[16]

6. 1 M Tris–HCl buffer, pH 7.5.

7. Hoechst 3358 solution: 1 µg/mL in distilled water.

2.4 Protein Synthesis Inhibitors (See Table 1)

1. Anisomycin solution (1,000×): 10 mg/mL (37 mM) anisomycin (Calbiochem) in 100 % ethanol. Store at –20 °C.

2. Emetine solution (1,000×): 25 mg/mL (45 mM) emetine dihydrochloride (Calbiochem) in 50 % ethanol. Store at –20 °C.

3. Harringtonine solution (1,000×): 2 mg/mL (3.7 mM) harringtonine (Santa Cruz) in 100 % ethanol. Store at –20 °C.

4. Puromycin solution (1,000×): 50 mg/mL (91 mM) puromycin (Calbiochem) in 50 % ethanol. Store at –20 °C.

5. Sodium arsenite solution (1,000×): 65 mg/mL (500 mM) sodium arsenite in distilled water. Store at 4 °C (*see* **Note 1**).

2.5 Antibodies

1. Anti-puromycin: we have tested three mouse monoclonal antibodies from different hybridoma clones, all of which are suitable for this protocol. Clone 12D10 was developed in P. Pierre's laboratory [17] and is commercially available (Millipore). Both 2A4 and 5B12 were developed in our laboratory, and 2A4 is available to the scientific community from the Developmental Studies Hybridoma Bank (http://dshb.biology.uiowa.edu/puromycin).

2. Anti-ribosomal P antibody: human polyclonal autoimmune antiserum which recognizes three proteins of the 60S ribosomal subunit (ImmunoVision, Springdale, AR, USA).

3. Anti-fibrillarin antibody: rabbit polyclonal (Abcam).

4. Donkey anti-mouse IgG (H + L) Alex Fluor 488 (Jackson ImmunoResearch).

5. Donkey anti-rabbit IgG (H + L) Alexa Fluor 594 (Jackson ImmunoResearch).

6. Donkey anti-human IgG (H + L) Cy5 (Jackson ImmunoResearch).

2.6 Equipment

1. High-quality glass coverslips, 12 mm diameter, #1 thickness (Assistent). Autoclave before use.

2. Microscope slides (Thermo).

3. Six-well and 24-well plates (Corning).

4. Microscopy forceps. Autoclave before use.

5. Laser scanning confocal microscope: we use a Leica TCS SP5 with an HCX PL APO lambda blue 63.0× 1.40 oil UV objective and LAS AF V2.3.1 software, Imaris (Bitplane) and Huygens Essential Software for image deconvolution using the classical maximum likelihood estimation algorithm (V3.6, Scientific Volume Imaging BV, Hilversum, The Netherlands). Other comparable systems may be used.

3 Methods

The method we describe here includes controls needed to ensure the specificity of both the labeling and the labeling steps, namely, three "translation inhibitor controls" and one "no puromycin" control. Moreover, we describe a quadruple labeling protocol that permits simultaneous visualization of nuclear translation sites (puromycin), large subunit ribosomal proteins (RPLP0, RPLP1, RPLP2), nucleoli (fibrillarin), and DNA (Hoechst 3358).

1. On day 1, pre-warm growth medium in a 37 °C water bath. Add 4 mL of warm growth medium per well to five wells of a 6-well plate. Then, carefully place up to five non-overlapping coverslips in each well using sterile forceps (*see* **Note 2**).

Transfer 1 mL of growth medium containing 0.5×10^6 HeLa cells per mL into each well. Incubate for 24 h to allow the cells to attach and stretch tightly to the coverslips.

2. On day 2, check the cells with an inverted microscope to ensure they have reached 70–90 % confluence. Warm 10 mL of growth medium, 4 mL of freshly prepared labeling medium, and 1 mL of labeling control medium to 37°. Meanwhile, chill 40 mL of PBS and 5 mL of freshly prepared extraction buffer on ice.

3. This step involves pre-incubation of cells with inhibitors in order to control the specificity of the puromycin labeling (*see* Table 1). Dilute each inhibitor in 2 mL of pre-warmed growth medium. This protocol is designed to accommodate three "translation inhibitor controls." Assign a well for each inhibitor, aspirate the medium, and replace it with medium containing the corresponding inhibitor. Incubation times are specified in Table 1 (*see* **Note 3**).

4. Prepare labeling medium: add 5 µL of puromycin stock solution (to 91 µM final concentration) and 5 µL of emetine stock solution (to 45 µM final concentration) to 5 mL of growth medium. Prepare labeling control medium: add 1 µL of emetine stock solution (45 µM final) to 1 mL of growth medium.

5. Aspirate the medium from each well and replace with 900 µL of pre-warmed labeling medium (test well + the 3 "inhibitor control" wells) or labeling control medium (in the last well). Incubate for 5 min at 37 °C (*see* **Note 4**).

6. Place the 6-well plate on ice, aspirate the medium, and replace it with 5 mL of ice-cold PBS.

7. Aspirate the PBS and add 1 mL of ice-cold extraction buffer to each well. Incubate for 5 min on ice (*see* **Note 5**).

8. Aspirate the extraction buffer and extremely gently add 1 mL of ice-cold wash buffer.

9. Gently aspirate the wash buffer and extremely gently add 1 mL of freshly made 3 % PFA solution (diluted in PBS) (*see* **Note 6**). Incubate for 15 min at room temperature and then replace the fixing solution with 2 mL of PBS. Make sure, by observing on an inverted microscope, that you have not lost all of the cells during the preceding steps.

10. Transfer one coverslip for each experimental condition into a 24-well plate, and save the others at 4 °C. Then incubate the cells with 500 µL of labeling buffer for 15 min at room temperature. Meanwhile, dilute the primary antibodies in labeling buffer: anti-puromycin to 1–4 µg/mL depending on the clone (we generally dilute our purified batch of 2A4 and 5B12 to 2 µg/mL), anti-ribosomal P to 1/5,000, and anti-fibrillarin to 1/200 (*see* **Note 7**).

11. Lay down a small piece of Parafilm, and spot 30 µL of diluted primary antibody on the Parafilm.

12. Using forceps, carefully remove a coverslip, remove excess labeling buffer by gently blotting its edge on absorbent tissue, and place cell-side down on the drop of primary antibody (*see* **Note 8**). Cover the samples with a Petri dish with a moist paper towel attached to the interior of the top and incubate at room temperature for 60 min (*see* **Note 9**).

13. Dilute the secondary antibodies in labeling buffer: 1/500 for goat anti-mouse Alexa Fluor 488, 1/500 for donkey anti-rabbit Alexa Fluor 594, 1/500 for donkey anti-human Cy5.

    Using forceps, remove coverslips from the drops of primary antibody and place them cell side up in a 24-well plate. Wash three times with 1 mL of PBS.

14. Spot five 30 µL drops of diluted secondary antibody solution on Parafilm and invert the coverslips on the drops as described in **step 13**, and place under a Petri dish lid covered with aluminum foil to protect the fluorophore from light. Incubate for 60 min.

15. Pick up the coverslips with forceps from the primary antibody drops and place them cell side up in a 24-well plate. Wash twice with 1 mL of PBS, then with 1 mL of distilled H_2O.

16. Aspirate the distilled H_2O and add 200 µL of Hoechst 3385 solution. Incubate for 5 min at room temperature.

17. Aspirate the solution and wash twice with distilled H_2O (*see* **Note 10**).

18. Place drops (5 µL) of Fluoromount-G on slides (*see* **Note 11**).

19. Using forceps, carefully pry up an edge of the coverslips, gently place them cell-side down on absorbent tissue to remove water, and then place them cell-side down on the drop of Fluoromount-G solution (*see* **Note 12**).

20. Place the slides in a tray and leave them to dry overnight, in the dark to protect the fluor from light (*see* **Note 13**). Store the tray in the refrigerator until analysis. Labeling is stable for at least 2 weeks at 4 °C in darkness. For longer storage we recommend using slide boxes and storing at –20 °C.

21. Examine the slides with a confocal microscope (Fig. 2) (*see* **Note 14**).

4   Notes

1. Sodium arsenite is harmful if inhaled and can be fatal if swallowed or absorbed through the skin.

2. Coverslips are easily breakable and accidents happen. Moreover, it is always interesting to associate RPM with multiple labeling (Fig. 2). For these reasons we usually work with four coverslips per well. To be certain that coverslips do not move around in

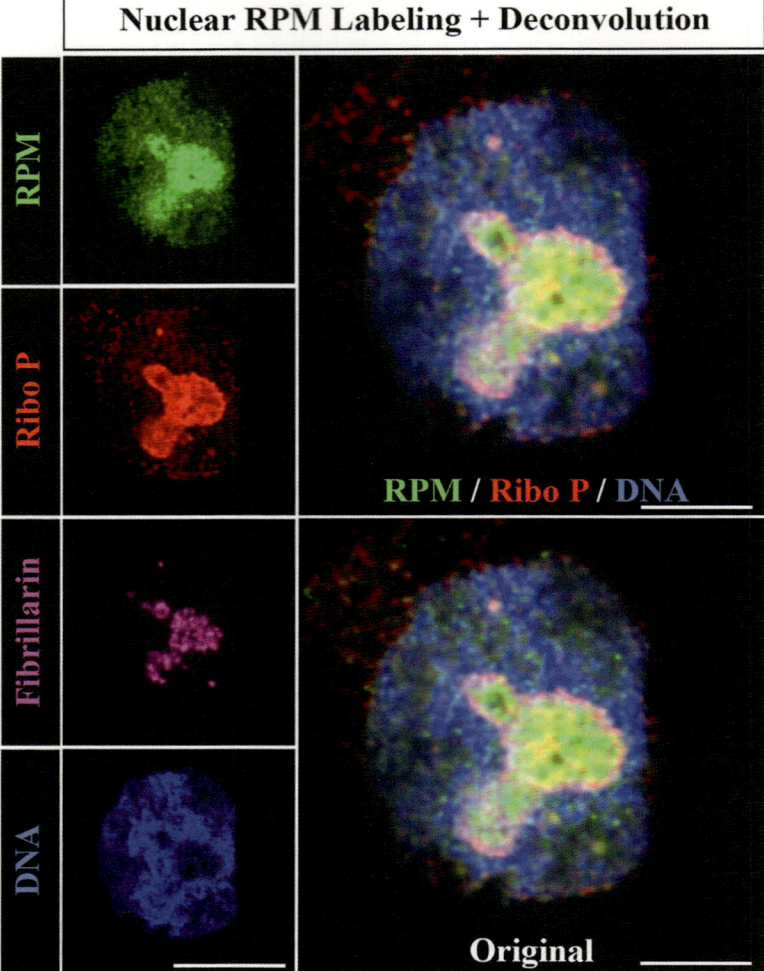

Fig. 2 High magnification series of a HeLa cell treated with the nuclear RPM protocol. The *left column* shows a single section from each probe. A merged image of this section is shown without (below) or with (above) deconvolution to minimize unfocused light from other focal planes. Scale bars = 5 μm. Reproduced from [7]

the well and overlap each other, gently press each coverslip in order to form a thin film of water between the glass and the plastic that creates an adhesive force. High-quality forceps with extremely sharp tips are recommended to enable easy removal of the coverslips from wells. Practice is required to minimize breaking coverslips.

3. We recommend performing the RPM procedure in association with at least three control conditions: without "active ribosomes" using a translation initiation inhibitor (such as harringtonine) that results in ribosome run-off from mRNA,

thus maximally releasing nascent chains; without "ribosome-catalyzed puromycin incorporation" using a puromycin competitor such as anisomycin; and in the "absence of antigen," i.e., without puromycin labeling.

4. Since anisomycin is a reversible competitor of puromycin, we recommend adding anisomycin to the corresponding well during the puromycin labeling incubation (to maintain a constant anisomycin concentration).

5. Be careful when pipetting to avoid detaching the cells. It is crucial to add the extraction buffer slowly down the side of the well.

6. Following fixation, cells may be kept for at least 7 days without noticeably affecting the quality of the RPM labeling.

7. Glycine will quench the fixative properties of PFA. Saponin will facilitate the accessibility of the antibody to some epitopes. FBS plays the role of blocking agent and decrease nonspecific binding of primary antibodies. To minimize nonspecific binding of secondary antibodies, we usually supplement the diluted primary antibodies with 5 % serum from the species used to generate the secondary antibody. However, this addition might not be necessary when using secondary antibodies highly cross-adsorbed against IgG from the other species.

8. If you inadvertently drop the coverslip, reidentify the cell side by holding it up to a light and look for the side with a white film corresponding to the cell side (this takes practice; if you are not sure gently scratch each side with forceps and the cell side will show a mark).

9. In order to limit the use of precious antibodies you can spot only 10 μL. In this case, you definitely need to incubate in a humid chamber to prevent the coverslips from drying out.

10. PBS does not solubilize Hoechst 3385. For this reason, we recommend using distilled H_2O for washes after this step.

11. Fluoromount-G solution is quite viscous when cold. To facilitate the pipetting of small volumes we usually warm up the solution at room temperature for 15 min before using. Cutting the end of the micropipette tip with a razor or scissor also helps.

12. It is greatly advantageous to minimize the number of slides that have to be manipulated, and with practice up to 8 coverslips can be placed on each slide. Give careful thought to the order of the coverslips on the slide and put the most important coverslips for comparison with each other as closely as possible. The most important antibodies should be visualized with colors that can be seen by eye.

13. If needed, drying may be hastened by incubating slides at 37 °C for 2–3 h. Never examine slides before the mounting

solution is dry, as this can damage extremely expensive oil immersion objectives.

14. Microscopy should be performed in the dark to accommodate the eyes and obtain maximize visual acuity. Generally, the focal plane has to be established only once for each slide, allowing rapid viewing of all coverslips on the same slide. It is important to form a general impression of the labeling of as many cells as possible, and this is by far most easily done by eye. Equally important is to write down your conclusions of the labeling during or immediately after viewing the slides.

Acknowledgements

J.W.Y. is generously supported by the Division of Intramural Research, National Institute of Allergy and Infectious Diseases.

References

1. Allfrey VG, Mirsky AE, Osawa S (1955) Protein synthesis in isolated cell nuclei. Nature 176:1042–1049

2. Allfrey VG (1954) Amino acid incorporation by isolated thymus nuclei. I. The role of des-oxyribonucleic acid in protein synthesis. Proc Natl Acad Sci U S A 40:881–885

3. Iborra FJ, Jackson DA, Cook PR (2001) Coupled transcription and translation within nuclei of mammalian cells. Science 293:1139–1142

4. Nathanson L, Xia T, Deutscher MP (2003) Nuclear protein synthesis: a re-evaluation. RNA 9:9–13

5. Dahlberg JE, Lund E, Goodwin EB (2003) Nuclear translation: what is the evidence? RNA 9:1–8

6. Dolan BP, Knowlton JJ, David A et al (2010) RNA polymerase II inhibitors dissociate antigenic peptide generation from normal viral protein synthesis: a role for nuclear translation in defective ribosomal product synthesis? J Immunol 185:6728–6733

7. David A, Dolan BP, Hickman HD et al (2012) Nuclear translation visualized by ribosome-bound nascent chain puromycylation. J Cell Biol 197:45–57

8. Pestka S (1971) Inhibitors of ribosome functions. Annu Rev Microbiol 25:487–562

9. David A, Bennink JR, Yewdell JW (2013) Emetine optimally facilitates nascent chain puromycylation and potentiates the ribopuromycylation method (RPM) applied to inert cells. Histochem Cell Biol 139:501–504

10. Graber TE, Hébert-Seropian S, Khoutorsky A et al (2013) Reactivation of stalled polyribosomes in synaptic plasticity. Proc Natl Acad Sci U S A 110:16205–16210

11. Al-Jubran K, Wen J, Abdullahi A et al (2013) Visualization of the joining of ribosomal subunits reveals the presence of 80S ribosomes in the nucleus. RNA 19:1669–1683

12. Apcher S, Millot G, Daskalogianni C et al (2013) Translation of pre-spliced RNAs in the nuclear compartment generates peptides for the MHC class I pathway. Proc Natl Acad Sci U S A 110:17951–17956

13. Jiménez A, Carrasco L, Vázquez D (1977) Enzymic and nonenzymic translocation by yeast polysomes. Site of action of a number of inhibitors. Biochemistry 16:4727–4730

14. Hansen JL, Moore PB, Steitz TA (2003) Structures of five antibiotics bound at the peptidyl transferase center of the large ribosomal subunit. J Mol Biol 330:1061–1075

15. Kedersha N, Chen S, Gilks N et al (2002) Evidence that ternary complex (eIF2-GTP-tRNAiMet) – deficient preinitiation complexes are core constituents of mammalian stress granules. Mol Biol Cell 1:195–210

16. Fresno M, Jimenez A, Vazquez D (1977) Inhibition of translation in eukaryotic systems by harringtonine. Eur J Biochem 72:323–330

17. Schmidt EK, Clavarino G, Ceppi M et al (2009) SUnSET, a nonradioactive monitor protein synthesis. Nat Methods 6:275–277

Part IV

The Intranuclear Milieu

Chapter 12

Targeted Nano Analysis of Water and Ions in the Nucleus Using Cryo-Correlative Microscopy

Frédérique Nolin, Dominique Ploton, Laurence Wortham, Pavel Tchelidze, Hélène Bobichon, Vincent Banchet, Nathalie Lalun, Christine Terryn, and Jean Michel

Abstract

The cell nucleus is a crowded volume in which the concentration of macromolecules is high. These macromolecules sequester most of the water molecules and ions which, together, are very important for stabilization and folding of proteins and nucleic acids. To better understand how the localization and quantity of water and ions vary with nuclear activity, it is necessary to study them simultaneously by using newly developed cell imaging approaches. Some years ago, we showed that dark-field cryo-Scanning Transmission Electron Microscopy (cryo-STEM) allows quantification of the mass percentages of water, dry matter, and elements (among which are ions) in freeze-dried ultrathin sections. To overcome the difficulty of clearly identifying nuclear subcompartments imaged by STEM in ultrathin cryo-sections, we developed a new cryo correlative light and STEM imaging procedure. This combines fluorescence imaging of nuclear GFP-tagged proteins to identify, within cryo ultrathin sections, regions of interest which are then analyzed by STEM for quantification of water and identification and quantification of ions. In this chapter we describe the new setup we have developed to perform this cryo-correlative light and STEM imaging approach, which allows a targeted nano analysis of water and ions in nuclear compartments.

Key words Nucleus, Water content, Ions, Correlative microscopy, STEM, Cryo-microscopy, Nano-analysis, EDXS

1 Introduction

In the nucleus, the concentration of nucleic acids and proteins ranges between 75 mg/ml in the nucleoplasm and 400 mg/ml in condensed chromatin [1]. The folding, molecular recognition and stabilization of nucleic acids and proteins are strongly influenced by water and ions, but despite their critical roles in the organization and function of macromolecules their localization and quantification in nuclear compartments have been rarely studied [2].

Ronald Hancock (ed.), *The Nucleus*, Methods in Molecular Biology, vol. 1228,
DOI 10.1007/978-1-4939-1680-1_12, © Springer Science+Business Media New York 2015

Some years ago we showed that dark-field cryo-Scanning Transmission Electron Microscopy (cryoSTEM) and Energy Dispersive X-ray spectrometry (EDXS) are able to quantify the mass percentages of water, dry matter, and elements in freeze-dried ultrathin sections of cryofixed cells [3, 4]. This approach avoids the use of chemical fixatives or counterstaining to prevent diffusion of water and ions between cell compartments, and in consequence most nuclear compartments are hardly recognized by electron microscopy due to their low contrast to electrons. One way to identify a nuclear compartment (chromatin or the nucleolus, for example) is to localize and image it by the fluorescence of a specific GFP-tagged protein before imaging by electron microscopy in exactly the same region of interest (ROI) of an ultrathin section. This approach, the so-called Correlative Light Electron Microscopy (CLEM), has been widely applied and was extended to the field of cryoimaging (cryoCLEM) to produce high resolution electron tomograms of microtubules, vesicles, and mitochondria [5–9].

In cryoCLEM, ultrathin cryosections are deposited on finder grids with alphanumeric coding which facilitates the precise localization of ROIs for successive light and electron microscope imaging. In classical methods [6] grids are first placed on the stage of a fluorescence microscope in a dedicated cryosetup which maintains them at around −170 °C, protects them from condensation of water vapor, and allows recording of images by fluorescence. In a second step, the grid is unmounted and transferred into the cryoholder of the electron microscope for imaging at high resolution.

However, this useful method suffers from an important drawback due to the need to transfer the grid from the optical to the EM cryoholder. This transfer can induce: (1) mechanical or thermal damage to the very sensitive cryosections, (2) bending and contamination of the grid, and (3) random rotation and translation of the grid which greatly limit the alignment of optical and electron microscopy images. To overcome these drawbacks, we have developed a completely new method to perform cryoCLEM in which the grid is always inserted in the EM cryoholder. The latter is first introduced into a homemade cryosetup for fluorescence imaging, and then inserted into the EM for ultrastructural analysis [10]. In this chapter we describe this procedure, which avoids grid transfer and greatly increases the precision of the alignment of two images of the same ROI performed successively by light and electron microscopy. This setup, used with a STEM equipped for EDXS, has allowed us to perform targeted multimodal quantification of water and elements within nuclear subcompartments, for example chromatin, identified by the presence of a specific GFP-tagged protein [10, 11].

2   Materials

The results presented here were obtained using the following materials and solutions.

2.1   Reagents, Chemicals, and Solutions

1. Cells: HeLa cells stably expressing histone H2B-GFP [12].
2. Culture medium: Dulbecco's Modified Eagle's Medium (DMEM) (GlutaMAX, Gibco) with 10 % (v/v) fetal bovine serum (FBS).
3. Culture flasks: 25 cm^2 plastic (Nunc).
4. 0.25 % trypsin-EDTA solution (Gibco).
5. Dulbecco's Phosphate-Buffered Saline (DPBS) 1× (GIBCO).
6. Liquid nitrogen (Linde Gas).
7. Ethane (gas) (Air Liquid).

2.2   Equipment

2.2.1   Cryomethods

1. Aluminium pins 9.5 mm long, 2 mm diameter with one extremity hand-trimmed to ~0.8 mm diameter; for freezing samples.
2. Cryoplunge CP3 semi-automatic plunge freezing instrument (Gatan, Munich, Germany).
3. Cryoultramicrotome UC6-FC6 (Leica).
4. Cryoknife, 35° diamond (Diatome, LFG Distribution, Lyon, France).
5. Cryosphere (Leica) (provides an environment of <10 % relative humidity to avoid contamination of sections by ice crystals).
6. CRION ionizer (Leica) (dissipates electrostatic charges).
7. Formvar-carbon coated indexed copper grids (London Finder grids: LF-135-Cu) (Agar Scientific).
8. EM cryo holder 626 with temperature controller (Gatan).
9. Cryostage: homemade (Fig. 1g). A modified Dewar (vacuum) flask whose lower part contains liquid nitrogen (LN) and is thermally connected to a copper support placed above the LN. This maintains a dry N_2 atmosphere in the upper half of the Dewar in which the EM cryoholder is introduced laterally. This setup maintains the temperature of the grid, as measured by the thermosensor of the cryoholder, at −171.6 ± 1.1 °C for at least 1.5 h without refilling the Dewar flask with LN.
10. CM30 TEM/STEM 100 kV (Fei Tecnai) or similar fitted with a post-viewing screen and a STEM Dark-Field/Bright-Field Detector (Gatan).
11. STEM Annular Dark Field Detector (Gatan).
12. Energy Dispersive X-ray (EDX) 30 mm^2 Si (Li) R-SUTW detector (EDAX, Mahwah, NJ, USA).

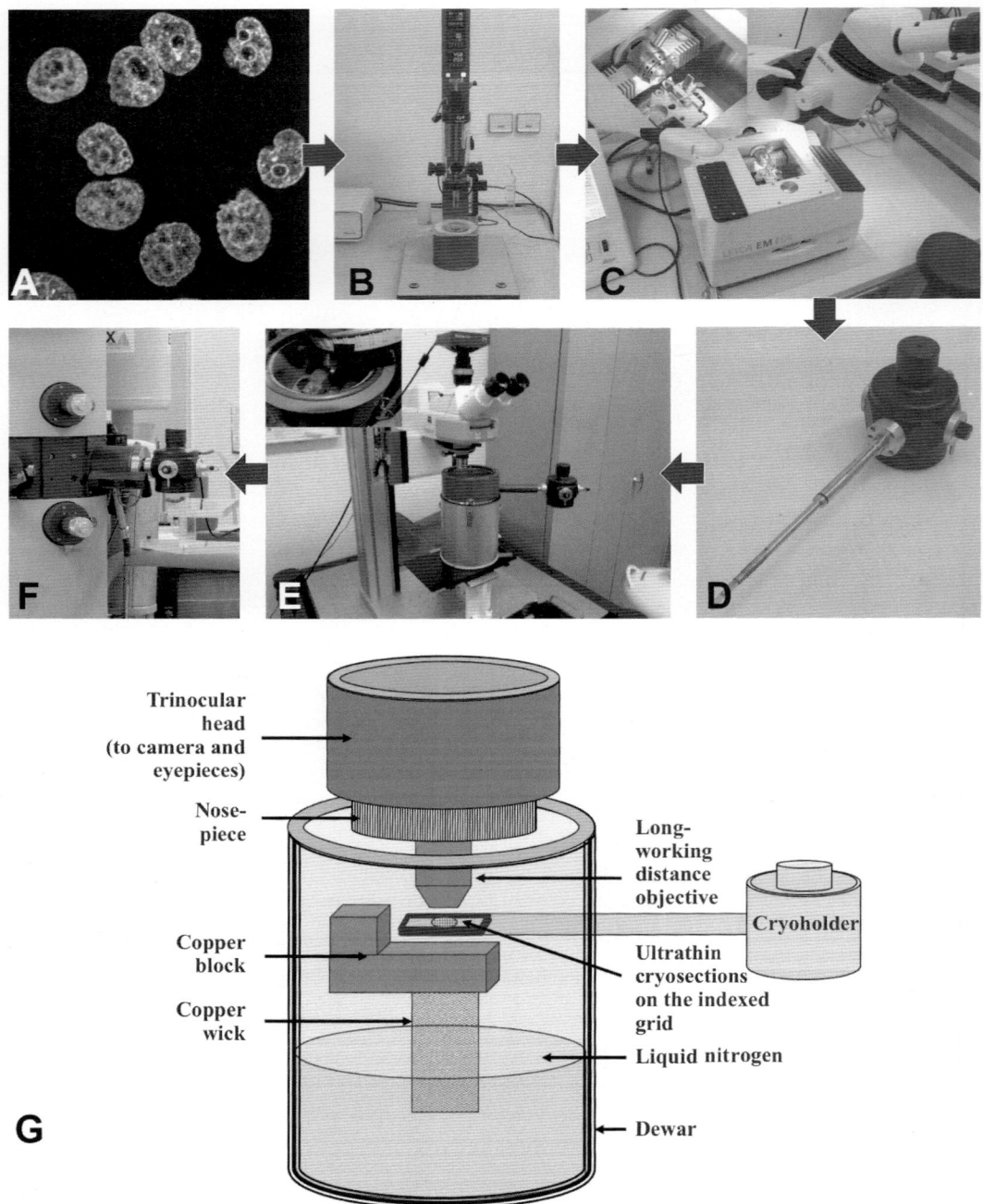

Fig. 1 Overview of the cryoCLEM procedure. (**a**) Fluorescence imaging of living HeLa cells expressing histone H2B-GFP to identify chromatin within the nuclei. (**b**) Cryofixation by plunging into liquid ethane. (**c**) Vitreous ultrathin sectioning. (**d**) Cryosections are placed on an indexed grid and inserted into the EM cryoholder. (**e**) Optical microscopy (fluorescence and reflected light) using a microscope with a large sample height clearance. The cryoholder is mounted on a homemade cryostage containing a LN_2 reservoir to keep a cold N_2 gas environment around the grid, maintaining the cryosections at a thermal steady-state of around −170 °C. (**f**) The cryoholder is then transferred into the electron microscope for freeze-drying of the cryosections before imaging and element analysis. (**g**) Schematic drawing showing the main parts of the homemade cryostage seen in (**e**). Panels **a**–**f** reproduced from [10] with permission

2.2.2 Optical Microscopy	1. Epifluorescence microscope: AxioScope Vario A1 with filters to detect GFP tagged-proteins and cold-light source with white, blue, and green LEDs for illumination (Zeiss) (*see* **Note 1**).
	2. CCD camera: AxioCam MRm (Zeiss).

2.3 Software

1. Axiovision (Zeiss).

2. DigitalMicrograph (Gatan).

3. ImageJ: available freely at http://rsbweb.nih.gov/ij/download. html.

4. Microanalysis X spectrum quantitation software (EDAX).

3 Methods

The procedure covers several stages: (1) preparation of cells for microscopy (optical and electron) by cryomethods; (2) fluorescence imaging with LED illumination for correlative microscopy using the EM cryoholder; (3) electron microscopy imaging using the same cryoholder; (4) dark field imaging for quantitation of water; (5) image processing for merging light and electron microscopy images; and (6) EDXS analysis for quantitating elements at the nanoscale. The protocol was elaborated for HeLa cells stably expressing histone H2B-GFP [12]. Cryomethods were chosen in order to avoid chemical fixation, dehydration, and embedding in plastic which modify the distribution of water and ions. Cryofixation by plunging into liquid ethane requires pellets of cells with a sufficient density to optimize the quality of cryosections, and we tested several conditions to obtain cell pellets which are suitable for reproducible structural and quantitative analysis.

3.1 Preparation of Cells

3.1.1 Cell Cultures

1. Maintain HeLa cells stably expressing histone H2B-GFP in DMEM-10 % FBS and reseed them in fresh medium as soon as monolayers reach confluence (2–3 times per week). Test cultures for mycoplasma infection.

2. Seed cells from a 48 h culture (80 % confluence) at 10^5 cells/mL in 25 cm^2 flasks.

3. Wash the cells with DPBS, cover them with 250 μL of trypsin solution, and incubate at 37 °C in 5 % CO_2 for 4 min. Resuspend the cells carefully in 350 μL of DMEM-FBS, transfer them to a 0.5 mL Eppendorf tube, and centrifuge at 37 °C to obtain a pellet.

3.1.2 Cryofixation (Fig. 1)

Cells are fixed by rapid plunging into liquid ethane to preserve the distribution of diffusible molecules in a state as close as possible to that in the living cell [10]. To prevent movements of water and ions, osmotically active cryoprotectants (sucrose or dextran) are not used. Frozen-hydrated samples are subsequently maintained in LN before vitreous sectioning.

1. In order to prepare for cryofixation of pellets, trim the extremity of an aluminium pin 9.5 mm long and 2 mm in diameter to ~0.8 mm diameter.

2. Pipette a drop (<1 μL) of the cell pellet (volume ~4 μL) on the trimmed extremity of the pin (*see* **Note 2**).

3. Hold the pin in the tweezers of a Cryoplunge and plunge it rapidly (1.7 m/s) into liquid ethane cooled by LN (*see* **Note 3**). After cryofixation, store the pin in a petri dish filled with LN until transfer into the cryoultramicrotome.

3.2 Cryoultramicro-tomy

1. Place the pin on the cryospecimen holder and lock this on the specimen arm of the cryoultramicrotome. Insert the diamond knife in the knife holder and mount the Cryosphere and CRION ionizer (*see* **Note 4**).

2. Cut vitreous sections with a nominal thickness of 85 nm and collect them on formvar-carbon-coated finder grids (*see* **Note 5**).

3.3 Cryotransfer of Grids for Fluorescence Microscopy

1. Place a grid with sections in the EM cryoholder maintained at −185/−190 °C by the temperature controller, and transfer the cryoholder to the homemade cryostage for fluorescence imaging (*see* **Note 6**).

3.4 Fluorescence Microscopy

1. Fix two fiducial markers on the stage of the microscope (one is seen in Fig. 1e as a red bar) to improve the alignment of images by light and electron microscopy. This allows precise positioning of the Dewar flask and limits its rotation to only a few degrees from one experiment to another.

2. Record black and white images. We use an EC Plan-Neofluar air objective (×10; 0.3 NA; 5.2 mm WD) for rapid screening of the entire grid and an LD LC Epiplan Neofluar air objective (×50; 0.55 NA; 9.1 mm WD) for imaging ROIs at higher resolution to identify fluorescent chromatin. The cold N_2 gas does not cause any problems for the objectives.

3. In a first step, image the center of the grid (marked with the two triangles) at low magnification using reflected light. Find the positions of the cryosections in reflected light and record images of H2B-GFP in fluorescence mode at low and high magnification (Fig. 2).

4. After imaging, close the shutter of the cryoholder to protect the sections during transfer to the electron microscope. Monitor the temperature of the grid continuously to confirm that cryopreservation is optimal during the whole procedure.

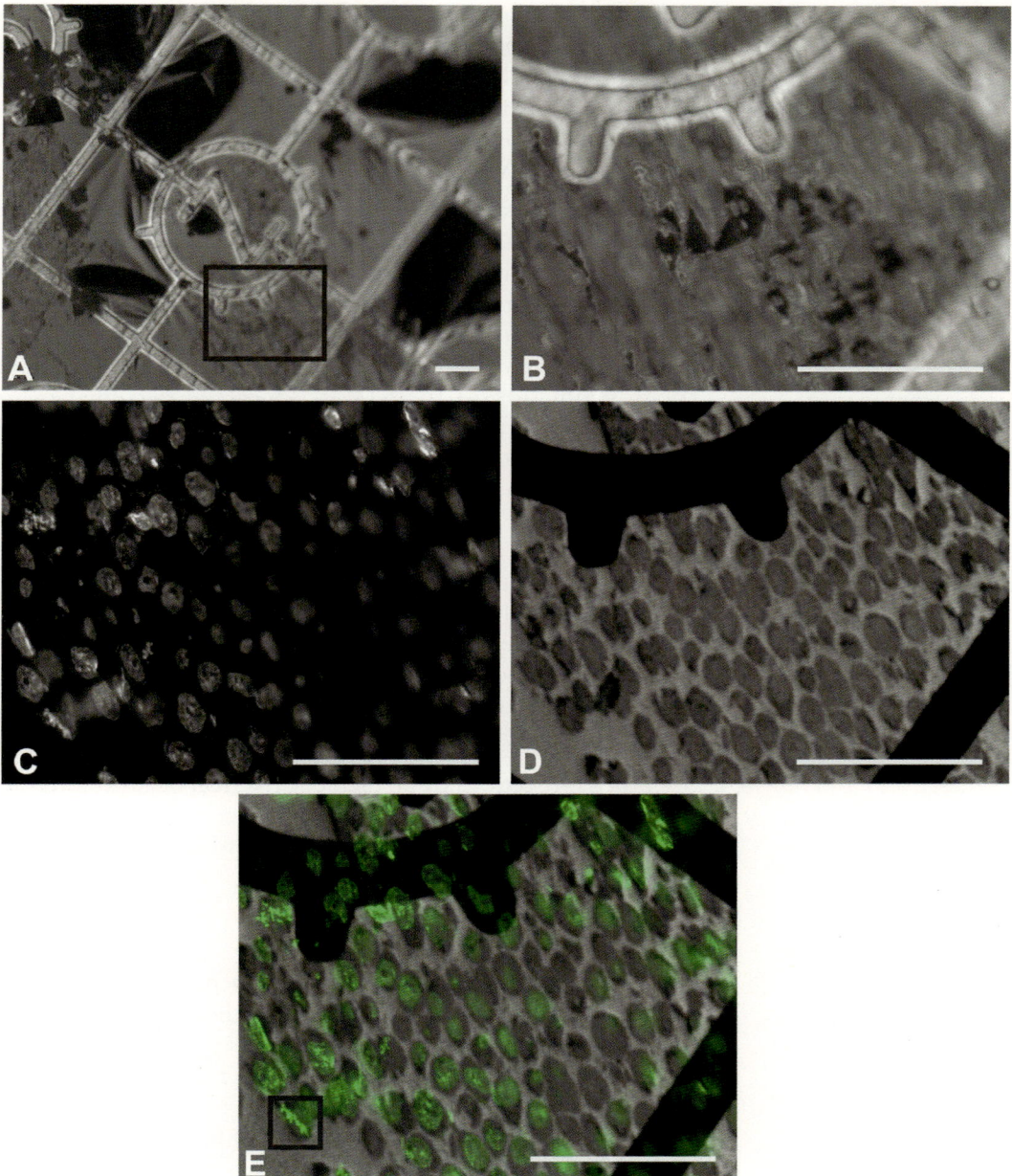

Fig. 2 CryoCLEM of ultrathin cryosections of cells collected on formvar and carbon coated indexed grids. (**a**) Reflected light image at low magnification showing grid bars, "Z" letter, and sections (×10, 0.3 NA objective). The *dark box* corresponds to the field of view in images (**b–e**). (**b**) Reflected light imaging at higher magnification (×50, 0.5 NA objective). (**c**) Fluorescence imaging of chromatin. Both mitotic cells and nuclei containing regions of condensed chromatin can be identified. (**d**) STEM bright field image after freeze-drying the same cryosection. (**e**) Superimposition of the fluorescence and STEM images obtained by merging images (**c**) and (**d**). The *square* in image (**e**) identifies a metaphase cell illustrated in Figs. 3d and 4. Scale bar 20 μm. Reproduced from [10] with permission

3.5 CryoSTEM Microscopy

1. Transfer the cryoholder into the electron microscope (*see* **Note 7**). Freeze-dry the grid in the microscope vacuum by slowly warming it up to –80 °C by increasing the warming current, using the temperature controller of the cryoholder. Cool the grid again and perform imaging at around –171 °C with an accelerating voltage of 100 kV.

2. Capture a low magnification (×500) image in STEM mode to find the two triangles marking the center of the grid and to place the grid in exactly the same position as in the light microscope. Compare this image with that made in reflected light mode on the same display screen, and rotate it precisely (for a few degrees if necessary) by using the image rotation function of the STEM. Once the two images are aligned, interesting fields previously identified by light microscopy are easily found in STEM mode thanks to their identical orientation.

3.6 Targeted Quantitation of Water

This dark-field STEM method is an indirect procedure using freeze-dried cryosections for measurements of water mass, first used by Zierold [13]. Its sensitivity is essentially limited by differences in the thickness of the cryosections, and we estimate an uncertainty of around 10 % in our experimental conditions which can be averaged by using a sufficient number of measurements for each biological condition.

1. Scan the cryosections by an incident electron beam at intensity I_0. A part of the elastically scattered electrons is collected by the annular dark-field detector. The recorded dark-field intensity is given by the equation:

$$I = I_0 \left(1 - \exp(-s\rho t)\right) \tag{1}$$

where s is the scattering parameter, dependent on the mean atomic number of the specimen and on the experimental conditions (accelerating voltage, angular acceptance of the detector); ρ is the sample mass density; and t is the local sample thickness. In our experimental conditions [10, 11] we can consider that $s\rho t \ll 1$, and Eq. 1 can be simplified as:

$$I = I_0 \left(s\rho t\right)$$

The dark-field intensity resulting from the film alone is $I_f = I_0 \left(s_f \rho_f t_f\right)$ and that resulting from the freeze-dried specimen alone is $I_s = I_0 \left(s_s \rho_s t_s\right)$. We can then define the relative dark-field intensity, i, by the equation:

$$i = \left(I_f + I_s\right) / I_f = 1 + \left(s_s \rho_s t_s / s_f \rho_f t_f\right) \tag{2}$$

We can consider the parameters s_f, ρ_f, and t_f to be constant if the support film is similar for different cryosections. Furthermore, the scattering parameter s_s is assumed to be constant due to the similarity of the scattering parameters for different species ρ_f biological macromolecules. We can also consider, as a first approximation, that the thickness t_s of freeze-dried cryosections remains constant under similar experimental conditions.

2. Quantify the water content (in %) for each pixel (20 nm in size) of the STEM image in a large number of cells analyzed in triplicate using DigitalMicrograph software. In the hydrated state, a cryosection consists of a dry mass portion D and a water mass portion L, with $D + L = 1$. The freeze-dried mass density ρ_s can therefore be written as $\rho_s =_D D$ where $_D$ is the mass density of the organic dry part which can be considered as a constant. Finally, Eq. 2 can be simplified as:

$$i = 1 + aD$$

where a is an experimental parameter to be determined using standard samples in similar experimental conditions. The water mass portion L can then be deduced directly from the relative dark-field intensity (I) measurements by:

$$L = 1 - \left((i-1) / a \right)$$

3. Compute parametric images for water content in which pixels containing 0–50 % water are shown in yellow and those containing 51–100 % water as a linear gradient from light to dark blue. Less hydrated cell compartments appear in yellow and light blue and more highly hydrated compartments in dark blue (Fig. 3).

3.7 Correlative Fluorescent and STEM Microscopy

Having shown previously that freeze-drying induces no differential shrinkage [10], the fluorescence and STEM images can be merged precisely by a simple procedure using ImageJ.

1. On the fluorescence image, crop a ROI of 150×150 pixels centred on the nucleus of interest and zoom it to the same size and magnification as the STEM image (512×512 pixels).

2. Adjust the brightness and contrast of the fluorescent image using the HiLo LUT command.

3. Draw the contour of the nucleus automatically using the threshold command. Use this contour to check or refine, if necessary, the zoom and orientation of the fluorescence image by merging it on the STEM image (image calculator command).

4. Apply rainbow LUT to the fluorescence image to transform grey levels into red, green and blue colours.

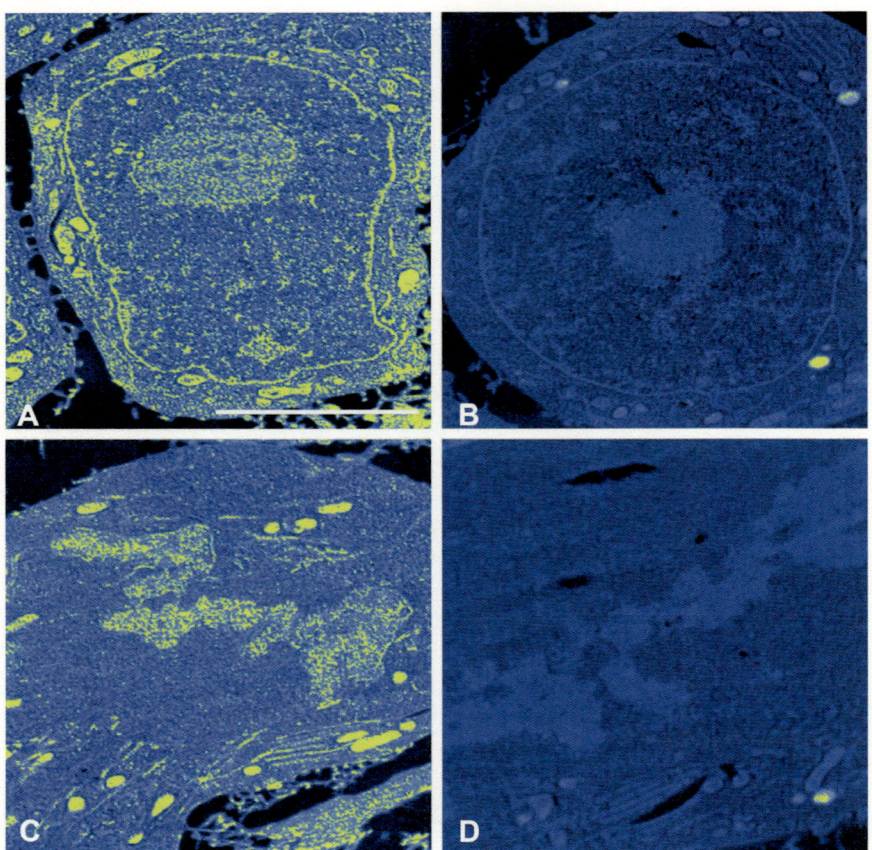

Fig. 3 Water content parametric images of control cells (*left column*) and of cells stressed by incubation with actinomycin D (AMD) (50 ng/mL) for 3 h after confluence (*right column*). Pixels containing 0–50 % water are shown in *yellow* and those containing 51–100 % water as a linear scale from *light* to *dark blue*. (**a**) Control cell in interphase. (**b**) AMD-treated cell in interphase. (**c**) Control cell in anaphase. (**d**) AMD-treated cell in meta-phase. The prevalent *dark blue* colour of AMD-treated cells illustrates their higher water content. The precise water content in the different cell compartments can be found in [11]. Scale bar 5 μm. Reproduced from [10] with permission

5. Merge this image (or those of each colour obtained by applying the split channel command) with the STEM image to create a composite image containing information from both fluorescence and STEM images (Fig. 4).

3.8 Targeted Quantitation of Ions

The above procedure is extremely useful to perform targeted nano analysis within structures identified by either their fluorescence pattern (such as uncondensed chromatin which is hardly identified due to its low contrast to electrons) or their intrinsic contrast and

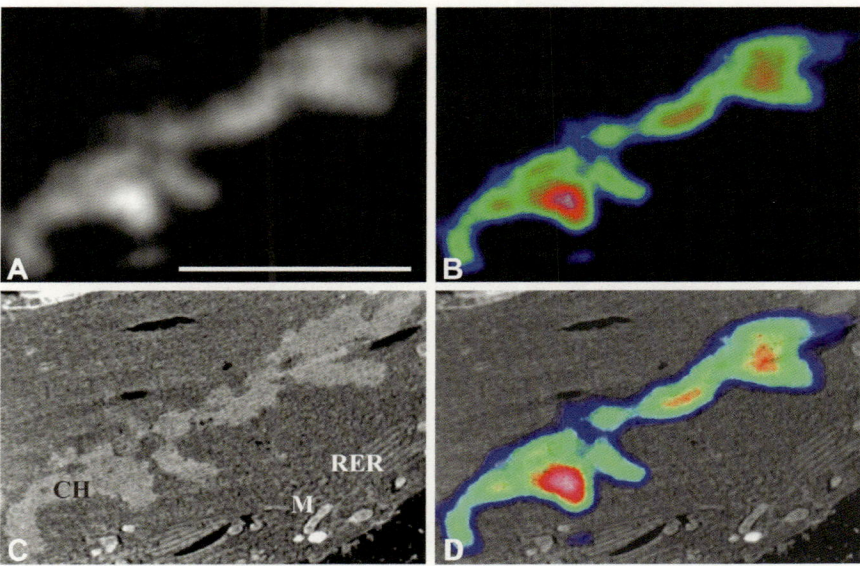

Fig. 4 CryoCLEM of a cell in metaphase. (**a**) Fluorescence image: metaphase chromosomes are clearly identifi-able within this ultrathin (85 nm thick) cryosection and show various fluorescence intensities. (**b**) Colour coding of high, medium, and low fluorescence intensities in *red*, *green*, and *blue*, respectively. (**c**) Dark-field STEM image after freeze-drying the same cryosection. This image shows the same metaphase cell with exactly the same orientation and field of view as in (**a**). Rough endoplasmic reticulum (RER) and mitochondria (M) can be identified in the cytoplasm due to their strong contrast to electrons. (**d**) Merge of images (**b**) and (**c**). Scale bar 5 μm. Reproduced from [10] with permission

shape (such as nucleoli and mitochondria). Light elements such as carbon, nitrogen, oxygen, sodium, potassium, and chloride are analyzed and quantified by EDXS.

1. Draw ROIs within compartments identified by their fluores-cence or their contrast and shape (Fig. 5).

2. Perform spectrum quantitation using homemade software according to Hall's continuum method and accounting for the signal from the support film and the spurious background signal from uncollimated electrons exciting the copper grid (*see* **Note 8**).

3. Convert the mass concentrations from mmol/kg of dry matter (C_D) into mmol/L (C_H) using the equation:

$$C_H = \left((100 - L) / L \right) C_D$$

where L is the percentage of water previously determined by quantitative STEM dark-field imaging.

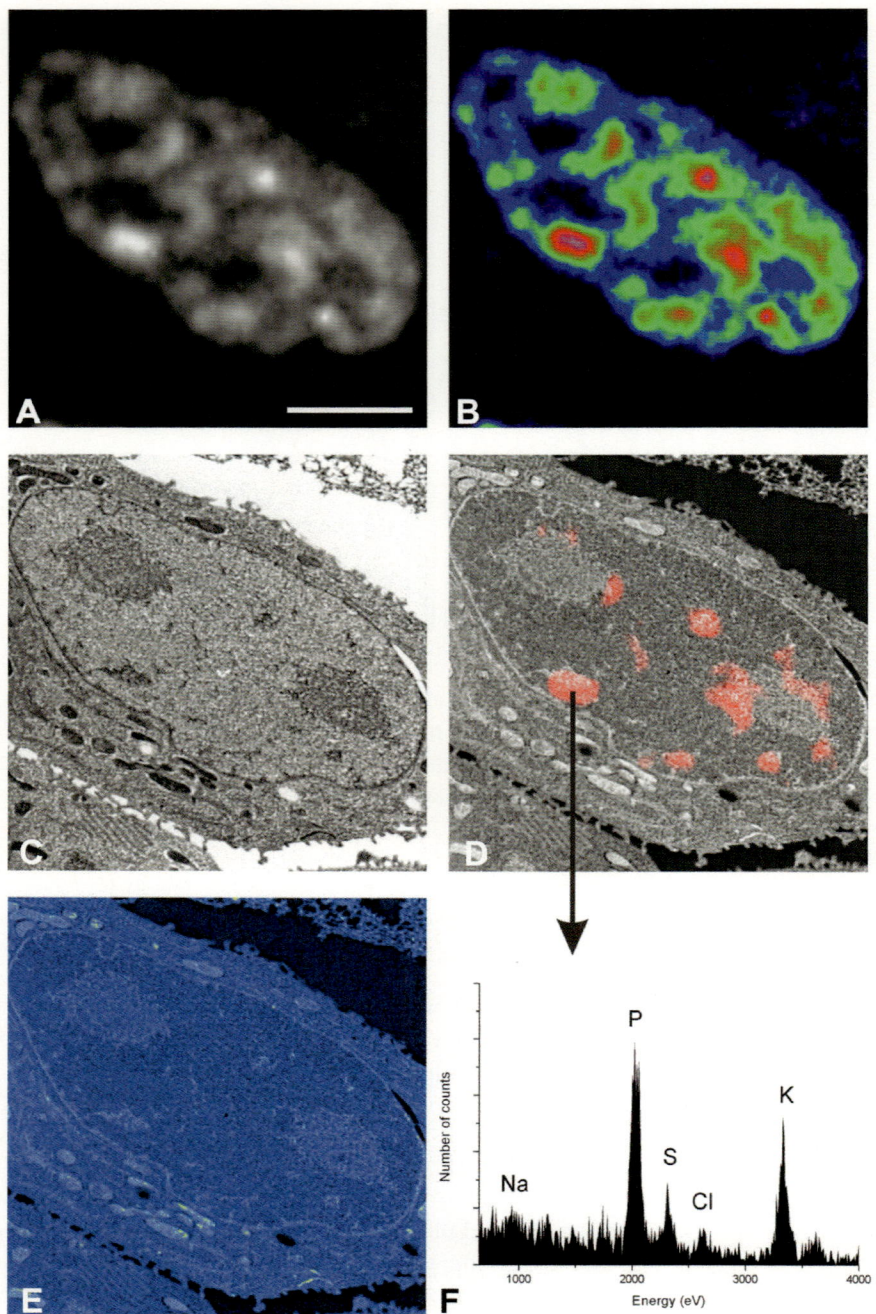

Fig. 5 Targeted nano analysis of water and ions in a nucleus. (**a**, **b**) Fluorescence mode: high, medium, and low fluorescence grey-scale levels in (**a**) are coded in *red*, *green*, and *blue*, respectively in (**b**). (**c**) STEM bright field image of the same freeze-dried cell. Note that the contrast of nucleoli and of some putative regions of chromatin is high. The contour of the nucleus and several mitochondria in the cytoplasm are easily recognized. (**d**, **f**) The image of the strongest fluorescence (*red* in **d**) identifying condensed chromatin was merged with the STEM dark-field image, allowing clear identification of condensed chromatin on the STEM image (**d**) and targeted EDXS analysis of one of these regions as seen in the recorded X-ray spectrum in (**f**). (**e**) Water content parametric image. Pixels corresponding to 0–50 % water are presented in *yellow* and those containing 51–100 % water are shown as a linear scale from light to *dark blue*. Scale bar 3 μm. Reproduced from [10] with permission

4   Notes

1. We use this microscope because its 56-cm tall column provides enough clearance to allow positioning the objective a few mm above the grid in the dry N_2 atmosphere (Fig. 1g). LEDs are used for illumination to prevent sample warming and frost formation on the grid even during long observation (1.5 h). A coverslip is not necessary to isolate the vitrified grid from the ambient atmosphere, which improves the image quality.

2. The viscosity of the pellet (related to cell density) is very important. The drop must stay on the extremity of the aluminium pin by a surface tension effect.

3. The gas used for the Cryoplunge CP3 is N_2, which is better than compressed air. Be careful with the gas pressure.

4. We use a 35° knife to avoid phenomena of compression and crevice formation. The CRION facilitates cryo section cutting and adherence and staggering over the grid. The Cryosphere provides good reproducibility of sectioning and permits cutting a large number of serial sections of the same cell.

5. Indexed EM grids allow rapid and efficient positioning of chosen ROIs for EM without the need for sophisticated software or desk motorization.

6. The cryochamber, knife, and cryospecimen holder temperature is –160 °C and the speed of cutting is 0.6 mm/s. The grid must be placed in its emplacement 5 min before to have time to cool.

7. We use the GATAN 626 cryoholder because it allows measurement and control of grid temperature. During cryotransfer of the grid all the equipment (forceps, specimen holder, etc.) must be precooled in LN_2. The temperature is maintained by the temperature controller at –185/–190 °C. The grid must be placed on the center of the specimen holder and not moved when the holding clips are added.

8. We tilt the specimen holder at 30° and do not analyze cells close to grid bars. We determined that the acquisition time necessary to record a spectrum with a good signal to noise ratio for most of the elements is 200 s. Advice on software for spectrum quantitation, which depends on the type of electron microscope used, is available from the authors.

Acknowledgments

We acknowledge funding from the Agence Nationale pour la Recherche (ANR-07Nano-CESIWIN), INSERM (Physicancer program: Noci-cytox), and Région Champagne Ardennes and technical support from the PICT IBiSA Biological Imaging Center from URCA University.

References

1. Hancock R (2014) The crowded nucleus. In: Hancock R, Kwang J (eds) New models of the cell nucleus: crowding, entropic forces, phase separation, and fractals, vol 307, Int Rev Cell Mol Biol., pp 15–26

2. Ball P (2008) Water as an active constituent in cell biology. Chem Rev 108:74–108

3. Terryn C, Michel J, Killian L et al (2000) Comparison of intracellular water content measurements by darkfield imaging and EELS in medium voltage TEM. Eur Phys J Appl Phys 11:215–226

4. Delavoie F, Molinari M, Milliot M et al (2009) Salmeterol restores secretory functions in cystic fibrosis airway submucosal gland serous cells. Am J Respir Cell Mol Biol 40: 388–397

5. Briegel A, Chen S, Koster AJ et al (2010) Correlated light and electron cryomicroscopy. Methods Enzymol 481:317–341

6. Sartori A, Gatz R, Beck F et al (2007) Correlative microscopy: bridging the gap between fluorescence light microscopy and cryoelectron tomography. J Struct Biol 160: 133–145

7. Schwartz CL, Sarbash V, Ataullakhanov F et al (2007) Cryofluorescence microscopy facilitates correlations between light and cryoelectron microscopy and reduces the rate of photobleaching. J Microscopy 227:98–109

8. Van Driel LF, Valentijn JA, Valentijn KM et al (2009) Tools for correlative cryofluorescence microscopy and cryoelectron tomography applied to whole mitochondria in human endothelial cells. Europ J Cell Biol 88:669–684

9. Rigort A, Bäuerlein FJB, Leis A et al (2010) Micromachining tools and correlative approaches for cellular cryoelectron tomography. J Struct Biol 172:169–179

10. Nolin F, Ploton D, Wortham L et al (2012) Targeted nano analysis of water and ions using cryocorrelative light and scanning transmission electron microscopy. J Struct Biol 180:352–361

11. Nolin F, Michel J, Wortham L et al (2013) Changes to cellular water and element content induced by nucleolar stress: investigation by a cryocorrelative nano-imaging approach. Cell Mol Life Sci 70:2383–2394

12. Kanda T, Kevin F, Sullivan KF et al (1998) Histone–GFP fusion protein enables sensitive analysis of chromosome dynamics in living mammalian cells. Curr Biol 8:377–385

13. Zierold K (1986) The determination of wet weight concentrations of elements in freeze-dried cryosections from biological cells. Scan Electron Microsc 2:713–724

Chapter 13

A Redox-Sensitive Yellow Fluorescent Protein Sensor for Monitoring Nuclear Glutathione Redox Dynamics

Agata Banach-Latapy, Michèle Dardalhon, and Meng-Er Huang

Abstract

Intracellular redox homeostasis is crucial for many cellular functions, but accurate measurements of cellular compartment-specific redox states remain technically challenging. Genetically encoded biosensors, including the glutathione-specific redox-sensitive yellow fluorescent protein (rxYFP), provide an alternative approach to overcome the limitations of conventional glutathione/glutathione disulfide (GSH/GSSG) redox measurements. In this chapter we describe methods to measure the nuclear rxYFP redox state in human cells by a redox Western blot technique. A nucleus-targeted rxYFP sensor can be used to sense nuclear steady-state and dynamic redox changes in response to oxidative stress. Complementary to existing redox sensors and conventional redox measurements, nucleus-targeted rxYFP sensors provide a novel tool for examining nuclear redox homeostasis in mammalian cells, permitting high-resolution readout of steady glutathione state and dynamics of redox changes. The technique described may be used with minimal variations to study the effects of stress conditions which lead to glutathione redox changes.

Key words Redox-sensitive biosensors, Redox protein extraction, Redox Western blot, Redox potential, Nuclear glutathione redox dynamics

1 Introduction

Measuring the nuclear glutathione/glutathione disulfide (GSH/GSSG) ratio and redox potential has long been extremely challenging due to the lack of adapted techniques. The development of redox-sensitive fluorescent proteins, such as redox-sensitive yellow fluorescent protein (rxYFP) [1] and reduction–oxidation green fluorescent protein (roGFP) [2] as glutathione redox reporters, has opened the way to measure the glutathione redox state in desired cellular localizations. The genetically introduced cysteine residues in rxYFP and roGFP equilibrate specifically with the GSH and GSSG redox pair via rapid thiol/disulfide exchange reactions through the action of glutaredoxins [3–5]. The ratio of oxidized to reduced rxYFP or roGFP measured in the cell can be used to generate an in vivo readout of the GSH/GSSG redox state.

Ronald Hancock (ed.), *The Nucleus*, Methods in Molecular Biology, vol. 1228,
DOI 10.1007/978-1-4939-1680-1_13, © Springer Science+Business Media New York 2015

The oxidation-dependent reversible formation of an intramolecular disulfide bond between two cysteine residues modifies their fluorescence properties and protein conformation. The roGFPs-based sensors have the advantage of exhibiting a ratiometric spectral shift, but require carefully calibrated microscopic measuring systems and general caution related to fluorescence artifacts and photobleaching. RxYFP is not ideal as a ratiometric probe, and is less suitable for fluorescence-based analysis compared to roGFPs. In contrast, the oxidized and reduced forms of rxYFP can be easily separated by nonreducing sodium dodecyl sulfate polyacrylamide gel electrophoresis (SDS-PAGE), providing a fluorescence-independent approach for redox analysis. This approach is particularly valuable to study ROS production and redox changes induced by pharmacological agents including anticancer drugs, as many of these display strong autofluorescence and require a long treatment time, complicating fluorescence-based analysis. We have recently targeted rxYFP to the nucleus of yeast cells as well as to the cytosol, mitochondrial matrix, and the nucleus of human cells to monitor compartment-specific glutathione redox state and its dynamics by redox Western blots [6, 7].

Here we describe the detailed protocol for using nucleus-targeted rxYFP to measure the nuclear glutathione redox state in human cells. We have constructed a nucleus-targeted rxYFP sensor that is expressed specifically throughout the nucleus [7]. The analysis of rxYFP redox state requires strict conditions of protein extraction that maintain the redox state of proteins. During cell breakage and protein extraction, reduced cysteine residues can undergo oxidation and conversely, oxidized residues can be reduced by thiol-disulfide exchange with cellular reductases, potentially making it difficult to evaluate the true in vivo redox state. Acidic quenching of thiol groups, which can circumvent this problem, is best achieved by treating cells with trichloroacetic acid (TCA, $pH < 1$) which also precipitates soluble cellular proteins (*see* **Note 1**). Reduced cysteine thiols are then blocked by the cysteine-specific alkylating agent *N*-ethylmaleimide (NEM) to prevent any redox modification. The reduced and oxidized forms of rxYFP protein are separated by non-reducing SDS-PAGE. Based on their relative quantities, the nuclear redox potential may then be calculated using the Nernst equation.

2   Materials

2.1   Cell Culture and Transfection

1. Cells: human cervical carcinoma (HeLa).

2. Growth medium: Dulbecco modified Eagle's medium (DMEM) supplemented with 10 % heat-inactivated fetal bovine serum (FBS), 2 mM L-glutamine, 100 µg/mL streptomycin, and 100 U/mL penicillin.

3. Tissue culture test plates with six wells (TPP, Trasadingen, Switzerland).

4. Plasmid based on the mammalian cell expression vector pCMV/cmyc/nuc (Life Technologies) expressing nucleus-targeted rxYFP (nuc-rxYFP) [7] (*see* **Note 2**). The plasmid is usually stored at a concentration of 200 ng/µL.

5. DNA transfection reagent: jetPRIME (PolyPlus, Strasbourg, France).

2.2 Redox Extraction

1. Trichloroacetic acid (TCA, Sigma-Aldrich): 15 % (w/v) solution in deionized water, store at 4 °C.

2. Acetone (Sigma-Aldrich): store at 4 °C.

3. Tris–EDTA–SDS buffer (TES): 100 mM Tris–HCl, pH 8.8, 10 mM EDTA, 1 % (w/v) sodium dodecyl sulfate (SDS), store at room temperature.

4. *N*-ethylmaleimide (NEM, Sigma-Aldrich): 1 M solution in ethanol, store at –20 °C and protected from light.

5. Quant-iT protein assay kit and Qubit fluorometer (Life Technologies).

6. Refrigerated microcentrifuge.

2.3 Redox Western Blot

1. NuPAGE LDS Sample Buffer 4× (Life Technologies).

2. NuPAGE Novex 12 % Bis-Tris Protein Gels, 1.0 mm × 10 wells (Life Technologies).

3. NuPAGE MOPS SDS Running Buffer 20× (Life Technologies).

4. SeeBlue Plus2 Pre-stained Protein Standards (Life Technologies) (*see* **Note 3**).

5. NuPAGE Transfer Buffer 20× (Life Technologies).

6. Pre-cut nitrocellulose membrane/filter paper sandwiches, 0.45 µm pore size, 8.3 × 7.3 cm (Life Technologies).

7. Phosphate-Buffered Saline (PBS): 210.0 mg KH_2PO_4, 9 g NaCl, 726.0 mg $Na_2HPO_4 \cdot 7H_2O$ per liter.

8. PBS containing 0.1 % (v/v) Tween-20.

9. Odyssey Blocking Buffer (Li-COR Biosciences, Lincoln, NE, USA): dilute 1:1 with PBS.

10. Rabbit anti-GFP polyclonal antibody (A11122, Life Technologies) (*see* **Note 4**).

11. IRDye 680RD- or IRDye 800CW-labeled goat anti-rabbit IgG secondary antibody (Li-COR Biosciences).

12. XCell SureLock Mini-Cell and XCell II Blot Module (Life Technologies) (*see* **Note 5**).

13. Odyssey Infrared scanner (Li-COR Biosciences).

3   Methods

3.1   Transfection

1. Seed ~2×10^5 HeLa cells in 2 mL of medium per well in 6-well plates (*see* **Note 6**).

2. Incubate the cells at 37 °C in a CO_2 incubator until they are 50–60 % confluent. This step usually takes 16–20 h.

3. Before transfection, prepare the DNA and transfection reagent in sterile tubes:

    (a) Dilute 1 μg of plasmid DNA in 200 μL of jetPRIME buffer for each well (*see* **Note 7**). Vortex briefly.

    (b) Add 2 μL of jetPRIME reagent (ratio 1 μg DNA:2 μL jetPRIME). Vortex for 10 s and spin down briefly.

    (c) Incubate the mixture for 10 min at room temperature.

4. Add 200 μL of this transfection mixture to the 2 mL of medium in each well. Mix gently. Incubate the cells at 37 °C in a CO_2 incubator.

5. Four hour after transfection, replace the medium with fresh normal medium.

6. Sixteen to 24 h after transfection, assess the transfection efficiency using an inverted fluorescence microscope equipped with an FITC filter (*see* **Note 8**).

3.2   Redox Protein Extraction

1. Protein extraction for measurement of the nuc-rxYFP steady redox state can be performed 24 h after transfection of the nuc-rxYFP-expressing plasmid. Cells at this stage are also ready for various experimental treatments to study dynamic nuc-rxYFP redox changes. For example, to analyze the time course of rxYFP oxidation following treatment with 100 μM H_2O_2, replace the medium with DMEM supplemented with 2 % FBS and 100 μM H_2O_2. Note the time of addition and stop the incubation at the appropriate time.

2. Aspirate the medium and wash the cells rapidly with PBS.

3. Overlay the cells with 1 mL/well of ice-cold 15 % TCA; make sure that the TCA covers all the cells. Keep the plates on ice for 30 min (*see* **Note 9**).

4. Remove the cells by scraping and transfer the solution containing the detached cells to 1.5-mL Eppendorf tubes. Keep the tubes on ice when not handled.

5. Pellet the cells by centrifugation for 20 min at full speed in a refrigerated microfuge (~$10,000 \times g$) at 4 °C.

6. Remove the TCA and wash the pellet with 200 μL of ice-cold acetone. Centrifuge for 5 min at ~$10,000 \times g$ at 4 °C.

Remove the acetone with caution to ensure recovery of the protein pellet. Repeat the acetone wash twice (*see* **Note 10**).

7. Remove the acetone carefully and dry the pellets under the hood for 10–20 min (*see* **Note 11**).

8. Add 100 μL of TES buffer containing 50 mM NEM to the pellet (*see* **Note 12**).

9. Incubate the samples for 1 h at 30 °C with vigorous shaking in the dark (*see* **Note 13**).

10. Clarify the lysates by centrifugation at ~10,000 × *g* for 10 min at room temperature.

11. Collect the supernatants and store them at –20 °C until ready for analysis or at –80 °C for longer storage. Avoid repeated freezing and thawing.

12. Determine the protein concentration in the extracts using a Quant-iT protein assay kit (*see* **Note 14**).

3.3 Redox Western Blot

1. Prepare an aliquot of the extract containing 30 μg of proteins in a total volume of 25 μL of LDS buffer without added reducing agents. Heat at 70 °C for 10 min. Spin briefly (*see* **Note 15**).

2. Prepare a protein gel and 1× electrophoresis running buffer. Set up the XCell SureLock Mini-Cell for electrophoresis.

3. Load molecular markers and samples into the wells of the gel. Electrophorese at a constant voltage of 150 V for at least 120 min for nuc-rxYFP (*see* **Note 16**). At the end of the run, remove the gel from the cassette for transfer.

4. Before performing the transfer, prepare 1× NuPAGE Transfer Buffer, soak the blotting pads in the transfer buffer for 1 h and remove air bubbles by squeezing them, place the pre-cut nitrocellulose membrane in transfer buffer for several min, and soak the pre-cut filter paper briefly in transfer buffer immediately before use (*see* **Note 17**).

5. Soak the gel in transfer buffer for a few minutes and rinse it with fresh transfer buffer before placing on the membrane (*see* **Note 17**).

6. Perform protein transfer using the XCell II Blot Module following the manufacturer's instructions. The gel is placed in the transfer sandwich (filter paper-gel-membrane-filter paper), cushioned by 3–5 pre-soaked blotting pads and held together to ensure complete contact of all components. Remove any trapped air bubbles during assembly of each layer.

7. Perform transfer at a constant voltage of 30 V for 1 h with the transfer unit on ice to dissipate heat produced during the run.

8. After transfer, place the membrane in PBS for few minutes with gentle shaking. Rinse twice with fresh PBS.

9. Block the membrane with Odyssey Blocking Buffer diluted 1:1 in PBS during 1 h at room temperature. Use at least 20 mL of blocking buffer/membrane (*see* **Note 18**).

10. Incubate the membrane with rabbit anti-GFP polyclonal antibody diluted 1:2,500 in Odyssey Blocking Buffer overnight at 4 °C with shaking (*see* **Note 18**).

11. Wash the membrane in PBS, 0.1 % Tween-20 for 5 min at room temperature and repeat this step twice more.

12. Incubate the membrane with IRDye-labeled goat anti-rabbit secondary antibody diluted 1:10,000 in PBS for at least 1 h at room temperature in the dark (*see* **Note 18**).

13. Wash the membrane in PBS, 0.1 % Tween-20 for 5 min at room temperature 3–4 times. Wash the membrane once in PBS and keep it in fresh PBS at 4 °C.

14. Scan the membrane using an Odyssey infrared scanner.

3.4 Scanning Procedure

Complete information for the scanning procedure can be found in the Odyssey User's Guide. A brief description of each scan parameter is given below:

1. Open the scanner and place the membrane facing down on the glass and on the lower left side of the grid.

2. Start the program by clicking on the "Odyssey" icon on the desktop.

3. In the "Scanner Console" window, name the scan and choose "Membrane" from the Preset list.

4. Specify the following scan parameters:

    (a) Resolution: 84 μm.

    (b) Quality: Medium.

    (c) Focus Offset: 0.

    (d) Channels: For IRDye 680RD-labeled secondary antibody, select channel 700 and uncheck channel 800. For IRDye 800CW-labeled antibody, select channel 800 and unselect channel 700.

    (e) Intensity: As a starting point, use 5.0. Rescan and adjust as needed.

    (f) Scan Area: Specify the scan area by pressing the left mouse button and dragging the cursor.

5. When the scan is complete, a reduced version of the image is shown on the scan grid. Click "Save" to save the scan.

3.5 Quantification of Results

1. Open the scanned image in Odyssey software. The image appears in colors.

2. Click "Add a new analysis to the selected scan" and Choose "Alter Image Display". In the pop-up window, select the

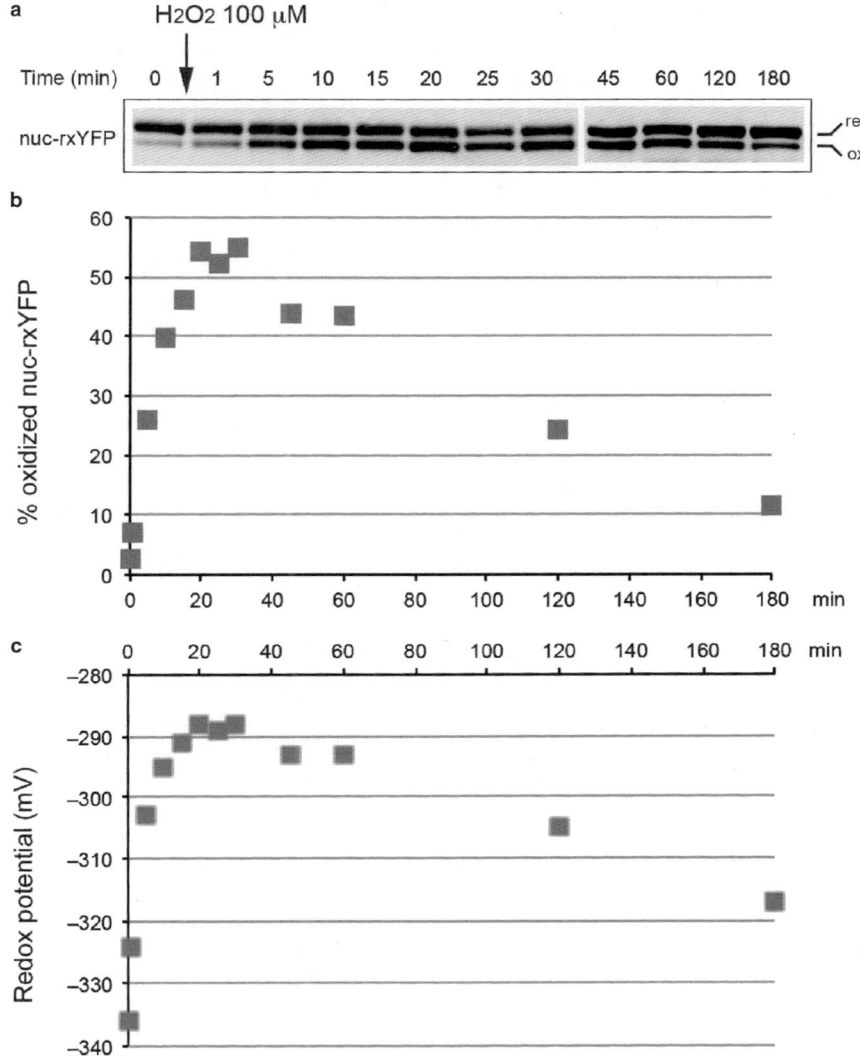

Fig. 1 (**a**) A representative redox Western blot showing the time course of oxidation of nuc-rxYFP in HeLa cells after addition of 100 μM H_2O_2 to the medium. (**b**) Quantification of nuc-rxYFP oxidation; each data point represents the mean value from three independent experiments. (**c**) Calculation of the nuclear glutathione redox potential using the Nernst equation. The values of nuc-rxYFP oxidation presented in (**b**) and the published value of pH 7.4 for the nucleus [9] are used for calculation of the redox potential

channel (700 or 800) according to the secondary antibody used (*see* Subheading 3.4, **step 4**d) and select "Show Gray Scale Image". Once the image appears in black and white, adjust the brightness and contrast to facilitate measurements (Fig. 1a).

3. Choose "Lanes" > "Add Multiple Lanes". In the pop-up window, choose the number of lanes to be analyzed and then select the area containing the bands of reduced and oxidized

rxYFP. Deselect all to enable working on separate lanes. For each lane, place the dashed line in the middle of the band and adjust the lane width according to the band size.

4. Define bands of interest by placing the rectangles on the reduced (upper) and oxidized (lower) bands separately and adjust their height to fully enclose the protein signal. Repeat for each lane (*see* **Note 19**).

5. Select all bands of interest and open "Background Method for Scan". Choose "Median" and "Right/Left". Avoid the "Top/Bottom" method, especially if the reduced and oxidized rxYFP bands are close (*see* **Note 20**).

6. Select all and create a report containing the Name of the band and its Integrated Intensity (II).

7. Calculate the percentage of oxidized protein relative to the total (Fig. 1b) using the equation:

$$\% \text{ oxidized} = \left[\text{II}_{ox} / \left(\text{II}_{ox} + \text{II}_{red} \right) \right] \times 100\%.$$

8. Calculate the glutathione redox potential E values (mV) (Fig. 1c). The glutathione redox potential E_{GSH} at 37 °C is calculated from the Nernst equation using the absolute concentrations of GSH and GSSG [8]:

$$E_{GSH} = E^{\circ'}{}_{GSH} - \left(61.5 / 2\right) \log \left(\left[GSH \right]^2 / \left[GSSG \right] \right)$$

where $E^{\circ'}{}_{GSH}$ is the standard potential for glutathione, which is –240 mV at pH 7. Because the redox potential of two redox couples that are in redox equilibrium is equal, the E_{GSH} can be calculated from the following equation using the ratio of reduced/oxidized rxYFP [4].

$$E_{GSH} = E^{\circ'}{}_{GSH} - \left(61.5 / 2\right) \log \left(\left[GSH \right]^2 / \left[GSSG \right] \right)$$
$$= E^{\circ'}{}_{rxYFP} - \left(61.5 / 2\right) \log \left(rxYFP_{red} / rxYFP_{ox} \right) = E_{rxYFP}$$

where $E^{\circ'}{}_{rxYFP}$ is the standard redox potential of rxYFP, which is –265 mV at pH 7 [4]. $E^{\circ'}{}_{GSH}$ or $E^{\circ'}{}_{rxYFP}$ must be adjusted for different pH conditions according to the following equation [8]:

$$E_{pH} = E^{\circ'} + \left[\left(pH - 7.0 \right) \times \left(-61.5 \right) \right]$$

where $E^{\circ'}$ is the standard redox potential of GSH or rxYFP at pH 7.0. Thus, the standard redox potential of rxYFP at pH 7.4

(the estimated nuclear pH value) [9] is −290 mV. The nuclear glutathione redox potential E_{nuc} can then be calculated (Fig. 1c):

$$E_{nuc} = -290\text{mV} - (61.5/2)\log(\text{rxYFP}_{red}/\text{rxYFP}_{ox})$$

4   Notes

1. The thiol group can engage in a redox reaction only when in the thiolated (deprotonated) state, which occurs more effectively when the pH of the solution is higher than the pKa of the cysteine residue. Cell treatment and lysis under acidic conditions with TCA (pH < 1) will keep cysteine residues in their protonated form, thus preventing any redox modifications.

2. The nucleus-targeted rxYFP plasmid is available on request from the corresponding author of this chapter (meng-er. huang@curie.fr).

3. The presence of a pink band of 19 kDa in the SeeBlue Plus2 Pre-stained Protein Standard is helpful to monitor the progression of protein migration in the gel and to control the run time. However, other protein standards may be used.

4. It is important to use an anti-GFP antibody that can recognize both oxidized and reduced rxYFP with similar efficiency. The rabbit anti-GFP polyclonal antibody (A11122) from Life Technologies displays such characteristics.

5. Other mini-gel electrophoresis and blotting apparatus can be used.

6. Seed a sufficient number of cells to reach 50–70 % confluence in 18–24 h. Seeding concentration varies depending on the use of rapidly or slowly growing cells.

7. Because of the size difference between nuc-rxYFP (~34 kDa) and cytosol-targeted rxYFP (cyto-rxYFP, ~29 kDa) or mitochondrial matrix-targeted rxYFP (mito-rxYFP, ~26 kDa), it is possible to co-transfect two plasmids (nuc-rxYFP and cyto-rxYFP, or nuc-rxYFP and mito-rxYFP) into the same cells [7]. Therefore, the redox state of two subcellular compartments can be analyzed simultaneously. In this case, the amount of each plasmid DNA could be reduced by half and the total DNA amount per well should not exceed that indicated in the protocol (1.0 μg).

8. Transfection efficiencies ranging between 60 and 80 % can usually be obtained in HeLa cells using the jetPRIME reagent. A transfection efficiency of >30 % is sufficient for detection of rxYFP signals in redox Western blots.

9. If needed, samples may be kept in 15 % TCA for several hours at 4 °C.

10. Avoid pipetting up and down and breaking the pellet with the tip. This does not improve washing efficacy, but causes loss of some amount of sample. As the pellet tends to detach from the tube wall after a wash, remove acetone with caution to ensure the recovery of the protein pellet.

11. Alternatively, pellets may be dried in a SpeedVac evaporator. Complete removal of acetone is necessary to enable the pellet to be dissolved in TES buffer. Dry pellets may be stored at −20 °C or at −80 °C for longer storage.

12. The volume of TES buffer depends on the volume of the pellet. To dissolve the pellet collected from one well of cells at ~80 % confluence, ~100 μL of buffer is needed. Adjust the volume of TES buffer according to the cell confluence and sample size to obtain similar protein concentrations and dissolve most proteins in the sample.

13. If the pellet is not completely dissolved after incubation, add more TES buffer to the sample and repeat the incubation.

14. The Quant-iT protein assay kit shows constant and reproducible quantification results for proteins in TES buffer compared to other techniques such as the Bradford method.

15. If the LDS buffer turns yellow, this may suggest that residual TCA is not completely removed from the protein extract, which changes the pH. The extract should be precipitated and re-washed.

16. For extracts from co-transfections (nuc-rxYFP and cyto-rxYFP, or nuc-rxYFP and mito-rxYFP), the gel may be run for ~120 min as for nuc-rxYFP alone. Follow the pink band in the SeeBlue Plus2 Pre-Stained Standard to determine the run time and stop the electrophoresis when it has migrated to ~5 mm above the bottom of the gel.

17. Prepare transfer buffer and materials while electrophoresis is in progress. Soak the gel, pads and membrane in separate trays. Removing air bubbles is essential as they can block the transfer of biomolecules. Soaking the gel in transfer buffer serves to remove running buffer.

18. As an alternative to Odyssey Blocking Buffer, 5 % (w/v) dry nonfat milk powder in PBS with 0.1 % Tween-20 may be used. Blocking buffer and both primary and secondary antibodies may be reused up to three times and stored for up to two weeks at 4 °C or at −20 °C for longer storage. Protect tubes containing the secondary antibody against light.

19. If extra bands in a lane are found by the software, remove undesired bands. If an expected band is not found by software, choose "Band" > "Add Band" and click on the area of interest.

20. If the background is uniform, choose "User Defined" as the background method. If the background is not uniform, choose the "Median" or "Average" method with "Right/Left".

Acknowledgments

This work was supported by the Centre National de la Recherche Scientifique (CNRS), the Institut Curie, and the Comité de l'Essonne de la Ligue Nationale Contre le Cancer (to MEH).

References

1. Ostergaard H, Henriksen A, Hansen FG et al (2001) Shedding light on disulfide bond formation: engineering a redox switch in green fluorescent protein. EMBO J 20:5853–5862

2. Hanson GT, Aggeler R, Oglesbee D et al (2004) Investigating mitochondrial redox potential with redox-sensitive green fluorescent protein indicators. J Biol Chem 279:13044–13053

3. Bjornberg O, Ostergaard H, Winther JR (2006) Measuring intracellular redox conditions using GFP-based sensors. Antioxid Redox Signal 8:354–361

4. Ostergaard H, Tachibana C, Winther JR (2004) Monitoring disulfide bond formation in the eukaryotic cytosol. J Cell Biol 166:337–345

5. Gutscher M, Pauleau AL, Marty L et al (2008) Real-time imaging of the intracellular glutathione redox potential. Nat Methods 5:553–559

6. Dardalhon M, Kumar C, Iraqui I et al (2012) Redox-sensitive YFP sensors monitor dynamic nuclear and cytosolic glutathione redox changes. Free Radic Biol Med 52:2254–2265

7. Banach-Latapy A, He T, Dardalhon M et al (2013) Redox-sensitive YFP sensors for monitoring dynamic compartment-specific glutathione redox state. Free Radic Biol Med 65:436–445

8. Schafer FQ, Buettner GR (2001) Redox environment of the cell as viewed through the redox state of the glutathione disulfide/glutathione couple. Free Radic Biol Med 30:1191–1212

9. Seksek O, Bolard J (1996) Nuclear pH gradient in mammalian cells revealed by laser microspectrofluorimetry. J Cell Sci 109:257–262

Part V

Imaging Nuclear Structures

Determination of the Dissociation Constant of the NFκB p50/p65 Heterodimer in Living Cells Using Fluorescence Cross-Correlation Spectroscopy

Manisha Tiwari and Masataka Kinjo

Abstract

Fluorescence cross-correlation spectroscopy (FCCS) is a promising technique for observing and quantifying protein–protein interactions in vitro and in vivo. FCCS has emerged as a useful tool for obtaining parameters of the concentration of labeled particles, their molecular dynamics, as well as the size of their complexes. This chapter discusses aspects of preparing a biological system for FCCS experiments and suggests practical advice for performing FCCS in living cells. Moreover, we describe the method of FCCS to determine the dissociation constant of a transcription factor dimer in the living cell.

Key words Fluorescence cross-correlation spectroscopy (FCCS), Laser scanning microscopy (LSM), Enhanced green fluorescent protein (EGFP), mCherry tandem dimer (mCherry$_2$), Immunoglobulin-like plexin transcription factor (IPT), Nuclear localization signal (NLS), Dissociation constant (K_d), Nuclear factor kappa B (NFκB), Heterodimer

1 Introduction

The NFκB p50/p65 heterodimer is the well-known transcription factor of the NFκB family. These proteins play an important role in regulating many genes such as those involved in inflammation, immune responses, and apoptosis [1, 2]. We have used fluorescence cross-correlation spectroscopy (FCCS) to quantify the heterodimerization of the IPT domain of p50/p65 in the living cell, and succeeded in observing interaction between the p50 and p65 proteins subunits [3]. In addition, the quantitative value of affinity, namely, the K_d, for the p50/p65 heterodimer in living cells, was estimated. The K_d values of mCherry$_2$- and EGFP-fused p50 and p65 were determined to be 0.46 μM in the cytoplasm and 1.06 μM in the nucleus of living cells. In this chapter, we present an approach using FCCS to determine the dissociation constant (K_d) of protein–protein interactions in the living cell.

Ronald Hancock (ed.), *The Nucleus*, Methods in Molecular Biology, vol. 1228,
DOI 10.1007/978-1-4939-1680-1_14, © Springer Science+Business Media New York 2015

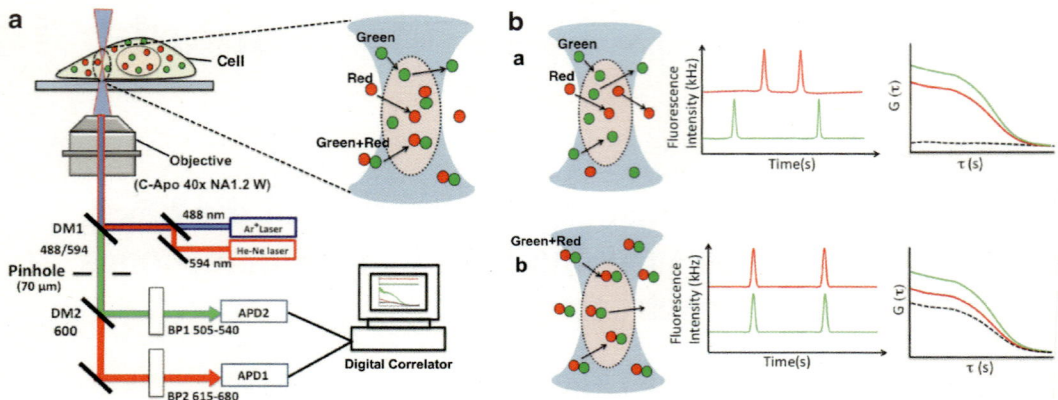

Fig. 1 Schematic experimental setup for fluorescence cross correlation spectroscopy (FCCS). (**a**) A small detection volume is defined by the optical system of the confocal microscope. Two spectrally different fluorophores are used. The fluctuation of fluorescence emission intensity originating from the movement of fluorescent molecules entering and leaving the confocal volume is monitored as function of time. Each fluorescent signal is simultaneously detected by each detector. The fluorescence fluctuation signals are transformed to an autocorrelation curve by the autocorrelation calculator unit, which also calculates cross-correlation functions between the two different fluorophores. (**b**) Schematic diagram of non-interacting (*a*) and interacting molecules (*b*) observed by FCCS. The *middle panel* represent the fluorescent intensity expressed as count rate (Hz, or number of photons per second). The *red* and *green lines* represent the red and green fluorescence, respectively. The *right panel* represents the auto and cross-correlation curve calculated using the fluctuation data of fluorescence intensity. The axis terms in the graph $G(\tau)$ and $\tau(s)$ represent the correlation function and the lag time, respectively. The *green*, *red* and *black curves* denote the autocorrelation of the green channel ($G_G(\tau)$), autocorrelation of the red channel ($G_R(\tau)$), and the cross-correlation curve ($G_C(\tau)$); respectively. A low cross-correlation amplitude is observed when the two differently labeled molecules are diffusing separately in the confocal volume and do not bind to each other (**a**). On the other hand, a high amplitude of cross-correlation is observed when the two differently labeled molecules are diffusing together through the confocal volume and are bound to each other (**b**)

FCCS is a versatile technique for observing direct associations between differently fluorescent-labeled proteins in a small sub-femtoliter confocal detection volume defined by the optical system of a confocal microscope, termed the diffraction-limited confocal volume (Fig. 1). The confocal volume (less than a femtoliter) allows us to resolve the measurement positions in the nucleus and cytoplasm. FCCS requires two fluorophores of different colors, and for simultaneous excitation of the two fluorophores using two lasers of different wavelengths and fluctuations in the fluorescence emission in this small volume are detected in two color detector channels. From the fluorescence signals detected in the two channels, two autocorrelation functions and one cross-correlation function are determined which contain information about the amount of free and interacting molecules. The parameters obtained using this method are the concentrations of the labeled particles (free and bound particles) and their diffusion constants, as well as the molecular sizes of their complexes [4]. This method is suitable for detecting interactions in a live cell environment quantitatively. FCCS has various applications to determine quantitative parameters, including determination of the dissociation constant (K_d) in the living cell [5–11]. Other than FCCS, there are few methods to determine the K_d values of biomolecules in living cells.

2 Materials

2.1 Cell Culture and Transfection

1. Cells: U-2 OS (human osteosarcoma) cells (ATCC HTB-96).

2. Culture medium: McCoy 5A modified medium supplemented with charcoal-stripped fetal bovine serum (Invitrogen) (10 % v/v), penicillin G (100 U/mL), and streptomycin sulfate (100 μg/mL).

3. Trypsin-EDTA solution (0.25 % v/v) (Sigma): store at −20 °C.

4. Optifect transfection reagent (Invitrogen).

5. OptiMEM I reduced-serum medium (Invitrogen).

6. Phosphate buffered saline (PBS): 137 mM NaCl, 2.7 mM KCl, 4.3 mM Na_2HPO_4, 1.47 mM KH_2PO_4, pH 7.4. Sterilize by autoclaving and store at room temperature.

7. 100 mm-diameter tissue culture dishes.

8. Chambered cover glasses for FCCS measurements: 8-well with No. 1 (0.17 mm) cover glass (Lab-TeK, Nunc, Rochester, NY).

9. Sterile 15 mL centrifuge tubes, 1.5 mL microtubes.

10. CO_2 incubator.

2.2 Equipment and Solutions for FCCS Analysis

1. LSM 510-ConfoCor 3 system with objective C-Apochromat 40×, 1.2 NA (Zeiss) (*see* **Note 1**).

2. Rhodamine 6G [R6G] fluorescent dye: 10^{-7} M aqueous solution.

3. Alexa Fluor 594 [Alexa 594] fluorescent dye: 10^{-9} M aqueous solution.

2.3 Plasmids

1. Plasmid encoding the IPT (immunoglobulin-like plexin transcription factor) domain of p65 fused with EGFP (p65-EGFP) and the IPT domain of p50 fused with a tandem dimer of mCherry (p50-mCherry$_2$).

2. For nuclear localization of the IPT domains of p50 and p65, the sequence encoding the SV40 large T antigen nuclear localization signal (Pro-Lys-Lys-Lys-Arg-Lys-Gly) is fused with the C-terminal end of mCherry$_2$ or EGFP. Subsequently, the IPT domains of p50 and p65 are inserted into the N-terminus of mCherry$_2$/NLS and EGFP/NLS, respectively (p50-mCherry$_2$/NLS, p65-EGFP/NLS) (*see* **Note 2** and Fig. 2).

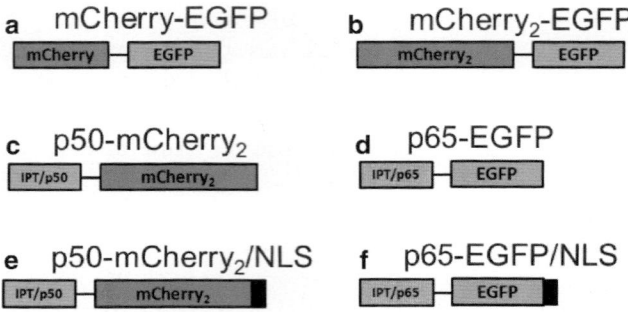

Fig. 2 Schematic diagrams of constructs. (**a**) mCherry fused with EGFP; (**b**) tandem dimer of mCherry (mCherry$_2$) fused with EGFP; (**c**) IPT domain of p50 fused with mCherry$_2$; (**d**) IPT domain of p65 fused with EGFP; (**e**) IPT domain of p50 fused with mCherry$_2$/NLS; (**f**) IPT domain of p65 fused with EGFP/NLS. *Black* regions indicate the NLS sequence of SV40

3. Plasmid encoding both mCherry$_2$ and EGFP (mCherry$_2$-EGFP) for positive control experiments (*see* **Note 3** and Fig. 4).

4. Plasmid encoding a tandem dimer of mCherry (mCherry$_2$), EGFP and mCherry$_2$/NLS, EGFP/NLS for negative control experiments (*see* **Note 4** and Figs. 3c and 7a).

3   Methods

3.1   Cell Culture

1. Seed cells in 100 mm culture dishes and grow in a CO_2 incubator (*see* **Note 5**).

2. Discard culture medium using an aspirator and wash the cells twice by adding 3 mL of PBS, swirling, and removing PBS with an aspirator.

3. Add 1 mL of 0.25 % trypsin solution, spread it over the monolayer of cells, and incubate for 3 min at 37 °C.

4. Add 10 mL of culture medium and transfer 1 mL of cell suspension to a 15 mL centrifuge tube.

5. Centrifuge the cell suspension at $130 \times g$ for 2 min.

6. Resuspend the cell pellet in 3 mL of culture medium.

7. Apply 300 µL of the cell suspension to each chamber of an 8-well chambered cover glass and place the cover.

8. Incubate overnight at 37 °C in a CO_2 incubator.

3.2   Transfection

1. Aspirate the medium from each well and gently replace with 300 µL of fresh culture medium.

2. Take new 1.5 mL microtubes, one for each well of cells, add 200 ng of p50-mCherry$_2$ or p50-mCherry$_2$/NLS, 100 ng of p65-EGFP or p65-EGFP/NLS, and 10 µL of OptiMEM.

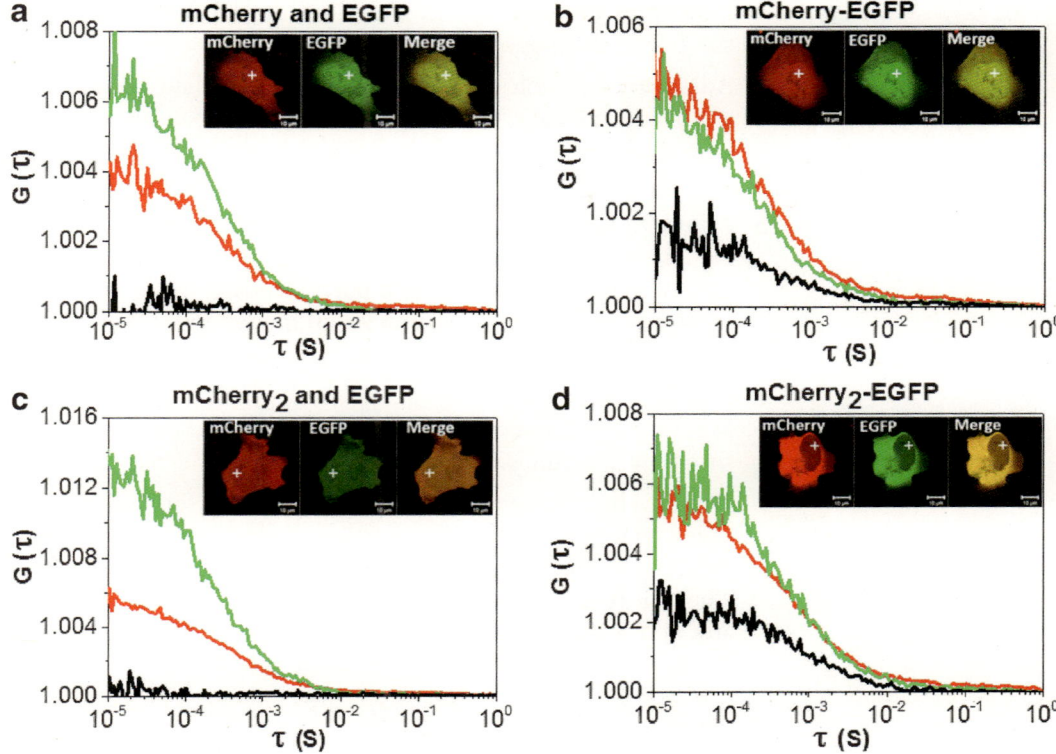

Fig. 3 FCCS measurement of mCherry-EGFP and mCherry$_2$-EGFP. Typical auto- and cross-correlation curves obtained from cells expressing a pair of fluorescent fusion proteins. The *green, red* and *black curves* denote the autocorrelation of the green ($G_G(\tau)$) and red channel ($G_R(\tau)$), and the cross-correlation curve ($G_C(\tau)$), respectively. The *inserts* show LSM images of cells expressing the fusion proteins with the FCCS measurement positions indicated by the white crosshairs. The scale bar represents 10 μm. FCCS measurement are shown of a cell coexpressing: (**a**) mCherry and EGFP (negative control); (**b**) mCherry-EGFP; (**c**) mCherry$_2$ and EGFP (negative control); (**d**) mCherry$_2$-EGFP

3. Prepare further new 1.5 mL microtubes, one for each well of cells. Add 1 μL of Optifect and 10 μL of OptiMEM to each tube, tap gently, and spin down. Incubate the tubes at room temperature for 5 min.

4. Combine the solutions from **steps 2** and **3** and incubate for 20 min at room temperature.

5. Add all the solution in each tube to each chamber of cells and incubate at 37 °C overnight in a CO$_2$ incubator (*see* **Note 6**).

3.3 Laser Scanning Microscope Imaging and FCCS Measurements

3.3.1 Confocal Adjustment

1. Adjust an appropriate optical pathway for the selected fluorescent dye.

2. Place 15–20 μL of solution containing 10^{-7} M Rhodamine 6G and 10^{-9} M Alexa Fluor 594 in a well of an 8-chambered cover glass.

3. Adjust the correction ring of the microscope lens to an appropriate counts/molecule setting. Adjust the *xy*-axis at the

fluorophore solution and the z-axis at 150–200 μm above the upper surface of the coverslip.

4. Adjust the pinhole size and its position by auto-adjustment according to the x, y and also z-axes by using fluorescent dyes (for example, R6G and Alexa 594). The pinhole size typically selected is equivalent to 1 Airy unit.

5. Adjust the excitation at 488 nm (Ar$^+$ laser) and 594 nm (He-Ne laser) to 15 and 8 μW respectively (*see* **Note 7**). The emission signals are split by a dichroic mirror (600 nm beam splitter) and detected at BP505-540 nm for the green channel and at BP615-680 nm for the red channel. Set the pinhole diameter at 70 μm.

6. Measure the fluorescence intensity for 30 s and repeat the measurement ten times.

7. Obtain the average autocorrelation curves for the red and green channels and a cross-correlation curve by using the algorithm described in the software package for the ConfoCor 3. The fluorescence autocorrelation functions from the green and red channels, $G_G(\tau)$, $G_R(\tau)$, and the fluorescence cross-correlation functions, $G_C(\tau)$, are calculated by:

$$G_G(\tau) = 1 + \frac{\langle \delta I_G(t) \cdot \delta I_G(t+\tau)\rangle}{\langle I_G(t)\rangle \cdot \langle I_G(t)\rangle} \tag{1}$$

$$G_R(\tau) = 1 + \frac{\langle \delta I_R(t) \cdot \delta I_R(t+\tau)\rangle}{\langle I_R(t)\rangle \cdot \langle I_R(t)\rangle} \tag{2}$$

$$G_C(\tau) = 1 + \frac{\langle \delta I_G(t) \cdot \delta I_R(t+\tau)\rangle}{\langle I_G(t)\rangle \cdot \langle I_R(t)\rangle} \tag{3}$$

where τ denotes the time delay, I_G is the fluorescence intensity of the green channel, I_R is the fluorescence intensity of the red channel, and $G_G(\tau)$, $G_R(\tau)$, and $G_C(\tau)$ denote the autocorrelation functions of green, red, and cross, respectively.

8. The $G(\tau)$ curves are fitted with the FCS fit program. After fitting the FCS data obtained with a three-dimensional single component model, determine the structure parameter. The detection volume element is defined by the structure parameter (s) representing the ratio of the beam waist (w_1) and the axial radius (w_2), $s = w_2/w_1$. The structure parameter for green and red channel is calibrated at this step. Detailed calculation of structure parameter values is discussed in Subheading 3.3.3.

3.3.2 Measurements

1. Observe the cells under the confocal microscope and select cells which express both green and red fluorescence with an appropriate intensity (*see* **Note 8**).

2. Scan the cells using a water immersion objective (C-Apochromat, 40×, 1.2NA).

3. Select measuring positions in the image (*see* **Note 9**).

4. Set the intensity of the excitation lasers to 1 % for the 488 nm and 0.5 % for the 594 nm, respectively (*see* **Note 10**).

5. Set the values of bleach time to 0 s, measuring time to 5 s, and repeat count to 10.

6. Measure the fluorescence intensity for each measuring position. Auto-correlation curves for red and green channels and a cross-correlation curve for red and green channels and a cross-correlation curve are obtained (Figs. 3, 6a, and 7a, b).

3.3.3 Curve Fitting and Data Analysis

The acquired auto- and cross-correlations are fitted using a two-component model as follows:

$$G(\tau) = 1 + \frac{1 - F_{triplet} + F_{triplet} \exp(-\tau / \tau_{triplet})}{N(1 - F_{triplet})}$$
$$\times \left(\left(\frac{F_{fast}}{1 + \tau / \tau_{fast}} \right) \sqrt{\frac{1}{1 + \tau / s^2 \tau_{fast}}} + \left(\frac{F_{slow}}{1 + \tau / \tau_{slow}} \right) \sqrt{\frac{1}{1 + \tau / s^2 \tau_{slow}}} \right) \quad (4)$$

where *Ftriplet* is the average fraction of triplet state molecules, *τtriplet* is the triplet relaxation time, F_{fast} and F_{slow} are the fractions of the fast and slow components, respectively, and τ_{fast} and τ_{slow} are the diffusion times of the fast and slow components, respectively. In the case of cross-correlation, fitting is performed as *Ftriplet*=0. *N* is the average number of fluorescent particles in the excitation-detection volume defined by radius (w_1) and half of the long axis (w_2) of the confocal volume element, and *s* is the structural parameter representing the ratio $s = w_2/w_1$. The values of w_1, i ($i = G$ or R) are determined from the diffusion coefficients of the rhodamine 6G and Alexa 594 used as standard dyes, respectively (*see* **Note 11**):

$$\omega_{1,i} = \sqrt{4D \cdot \tau_{Di}} \quad (5)$$

The volume elements *V* are calculated according to:

$$V_i = \pi^{3/2} \cdot \omega_{1,i}^2 \cdot \omega_{2,i} \quad (6)$$

$$V_C = \left(\frac{\pi}{2} \right)^{3/2} (\omega_{1,G}^2 + \omega_{1,R}^2)(\omega_{2,G}^2 + \omega_{2,R}^2)^{1/2} \quad (7)$$

The average numbers of green fluorescent particles (N_G), red fluorescent particles (N_R), and particles that have both green and red fluorescence (N_C) are given by:

$$N_G = \frac{1}{G_G(0) - 1} \quad (8)$$

$$N_{\mathrm{R}} = \frac{1}{G_{\mathrm{R}}(0) - 1} \tag{9}$$

$$N_{\mathrm{C}} = \frac{G_{\mathrm{C}}(0) - 1}{\left(G_{\mathrm{R}}(0) - 1\right) \cdot \left(G_{\mathrm{G}}(0) - 1\right)} \tag{10}$$

To subtract the effect of autofluorescence on N, corrected N ($N_{i,\,\mathrm{corrected}}$) is calculated by the following equation:

$$N_{i,\mathrm{corrected}} = N_{i,\mathrm{measured}} \cdot \left[1 - \frac{I_{i,\mathrm{background}}}{I_{i,\mathrm{measured}}}\right]^{2} \tag{11}$$

where $N_{i,\,\mathrm{measured}}$ is the average number of green or red fluorescent particles obtained from FCCS measurement and fitting analysis ($i = \mathrm{G}$ or R). $I_{i,\mathrm{measured}}$ is the average intensity of green or red fluorescence during measurement of FCCS ($i = \mathrm{G}$ or R). $I_{i,\,\mathrm{background}}$ is the average intensity of green or red fluorescence obtained from FCCS measurement of mock-transfected cells. Applying the corrected numbers for green and red to Eq. 11, the corrected number of cross correlated particles is calculated by the following equation (*see* **Note 12**):

$$N_{\mathrm{C},\mathrm{corrected}} = \left(G_{\mathrm{C}}(0) - 1\right) \cdot N_{\mathrm{G},\mathrm{corrected}} \cdot N_{\mathrm{R},\mathrm{corrected}} \tag{12}$$

- The concentration of each fluorescent protein is calculated with the use of A (Avogadro's number) as given below:

$$C_{i,\mathrm{corrected}} = \frac{N_{i,\mathrm{corrected}}}{V_i \cdot A} \tag{13}$$

$$C_{\mathrm{C},\mathrm{corrected}} = \frac{N_{\mathrm{C},\mathrm{corrected}}}{V_{\mathrm{C}} \cdot A} \tag{14}$$

- Diffusion constants of the samples are calculated from the ratio of the diffusion constant of Rh6G D_{Rh6G} ($414\ \mu\mathrm{m}^2/\mathrm{s}$) [15] and diffusion times τ_{R6G} and sample [16] (*see* **Note 13** and Figs. 6c and 7d).

3.3.4 Determination of K_d

The dissociation constant K_{d} is determined using following equations:

$$K_{\mathrm{d}} = \frac{[G_{\mathrm{free}}][R_{\mathrm{free}}]}{[\mathrm{Complex}]} \tag{15}$$

$$[G_{\mathrm{free}}] = [C_{\mathrm{G},\mathrm{corrected}}] - [C_{\mathrm{C},\mathrm{corrected}}] \tag{16}$$

$$[R_{\mathrm{free}}] = [C_{\mathrm{R},\mathrm{corrected}}] - [C_{\mathrm{C},\mathrm{corrected}}] \tag{17}$$

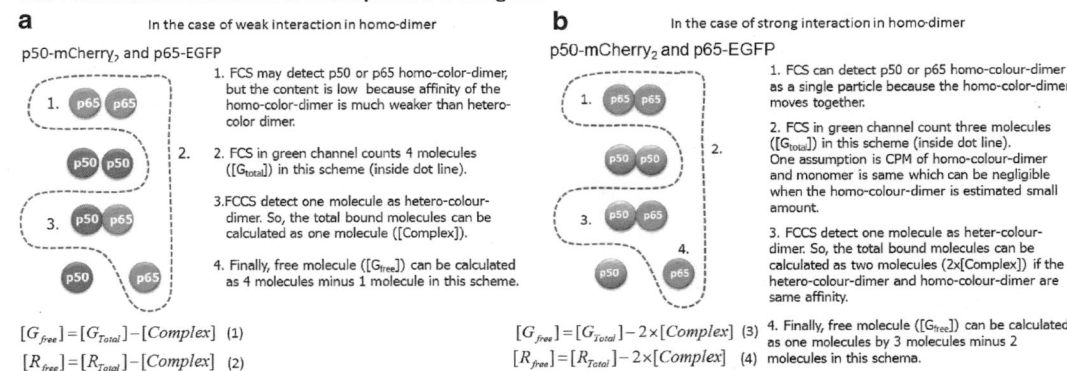

Fig. 4 Two possibilities of interaction of fusion proteins in living cells. (**a**) FCCS detects one molecule as a hetero-color-dimer in the case where the affinities of p50/p50 and p65/p65 homodimers are lower than that of p50/p65 heterodimer (unpublished data). In the case of the p50/p65 heterodimer, the total bound molecules can be calculated as one molecule [Complex] (Eqs. 1 and 2). (**b**) If the hetero-color-dimer and homo-color-dimer have the same affinity (Eqs. 3 and 4), the total bound molecules can be calculated as two molecules ($2\times$[Complex])

$$[\text{Complex}] = \left[C_{\text{C,corrected}}\right] \qquad (18)$$

The concentrations of the unbound EGFP and mCherry$_2$ fusion proteins $[G_{\text{free}}]$ and $[R_{\text{free}}]$ are calculated by subtraction of the concentration of the complex [Complex] from the total concentration of the EGFP and mCherry$_2$ fusion proteins (Fig. 4).

For exclusion of the background of cross-correlation and weak fluorescent signal, which mainly originates from fluorescence cross-talk between the two detectors at low concentrations of complex, the data points of less than the average concentration of "Complex" obtained from the negative control are excluded (Fig. 5). Then a scatter plot of the products of concentrations of free molecules versus the concentration of the complex is generated with a line of best fit and the dissociation constant (K_{d}) is calculated from the slope of the regression line [5] (*see* **Note 14** and Figs. 6b and 7c).

4   Notes

1. This protocol is based on the ConfoCor3 microscope system. In this system, a C-Apochromat objective (40×) with 1.2 numerical aperture and water immersion is recommended. The apparatus provides more flexible laser power tuning for each fluorescent probe, which gives us adequate fluorescence intensity to reduce pseudo-positive cross-correlation signals that can be caused by the tail of the fluorescence spectrum of EGFP [12]. The technical importance of this method is that flexible laser beam power using two lasers can be applied for

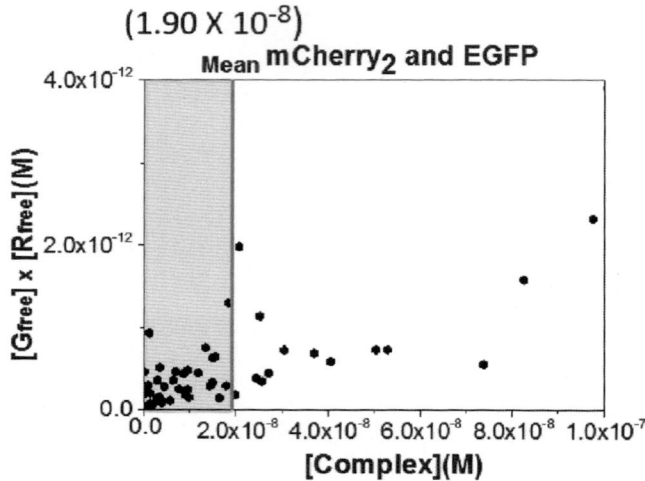

Fig. 5 Scatter plot of a negative control (independently expressed mCherry$_2$ and EGFP) showing the mean and exclusion area of data points. The plot represents the concentration of free mCherry$_2$ and of EGFP versus the concentration of the complex of mCherry$_2$ and EGFP. The red line shows the mean value (1.90×10^{-8}) of the concentration of complexes and the pink area represents the exclusion area of data points

any combination of proteins, homodimers or heterodimers for analysis of protein dynamics in living cells. All the Figures in this chapter are based on our previous results [3].

2. To examine protein–protein interactions, each target protein should be labelled by a spectrally different fluorescent protein, for example enhanced GFP (EGFP) for green fluorescence, monomeric RFP (mRFP) tandem dimer and mCherry for red fluorescence [3–12]. The constructs should be designed to protect the functionality of the target protein. Since the IPT domain of p50 and p65 subunits responsible for dimerization and DNA binding are not able to translocate into the nucleus, a nuclear localization signal is fused. The plasmids used here are described in [3] and are available from the Corresponding Author.

3. To examine the maximum cross-correlation, we compared the cross-correlation between monomeric mCherry fused with EGFP (mCherry-EGFP) and tandem dimer of mCherry fused with EGFP (mCherry$_2$-EGFP). A high cross-correlation amplitude was observed in mCherry$_2$-EGFP compare to mCherry-EGFP. mCherry$_2$-EGFP is used as a positive control.

4. To avoid the unknown interaction of any fused protein with fluorescent protein, mCherry and EGFP (and mCherry$_2$ and EGFP) is used as negative control.

5. In this case, interaction of p50 and p65 was observed. If necessary, select the cell line depending on the study. For calculation

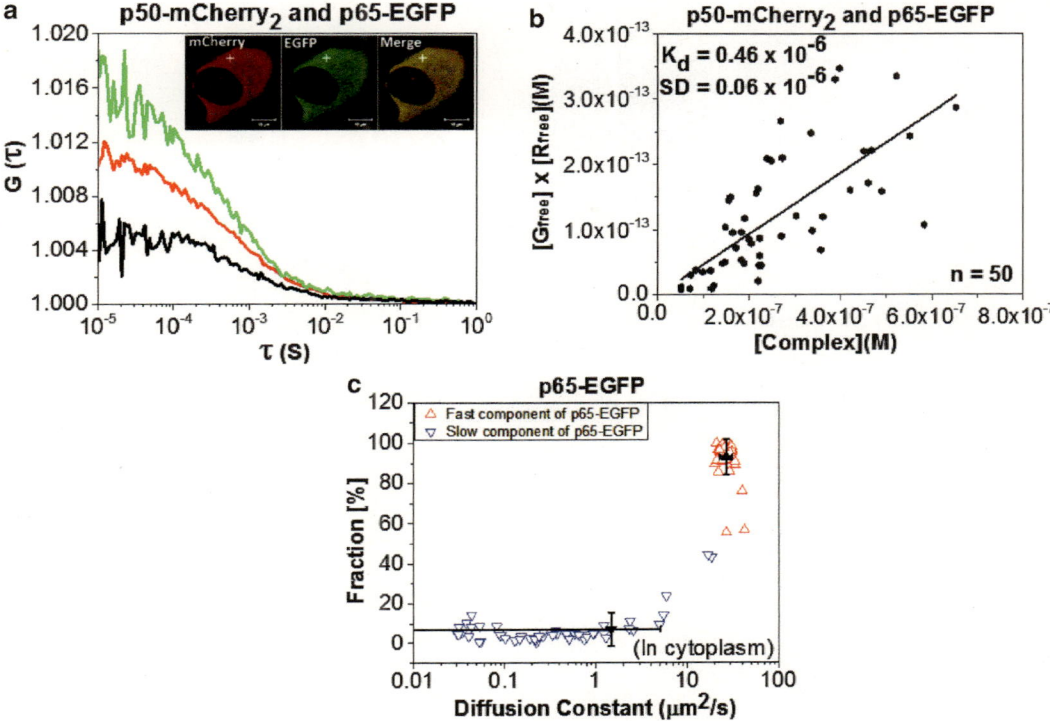

Fig. 6 FCCS measurement and K_d analysis of differently-labeled p50 and p65 in the cytoplasm of living cells. (**a**) Typical auto and cross-correlation curves from cells coexpressing the pair p50-mCherry$_2$ and p65-EGFP in the cytoplasm. The *green, red* and *black curves* denote the autocorrelation of the green channel ($G_G(\tau)$), the autocorrelation of the red channel ($G_R(\tau)$), and the cross-correlation curve ($G_C(\tau)$), respectively. The *insert* shows LSM images of cells coexpressing the pair p50-mCherry$_2$ and p65-EGFP with FCCS measurement positions indicated by the white crosshairs. The scale bar represents 10 μm. (**b**) Results of K_d determination using a scatter plot and linear regression. The plot represents the concentration of free mCherry$_2$ and EGFP fusion proteins versus the concentration of the complex of mCherry$_2$ and EGFP fusion proteins. The *solid line* shows a linear fit. The *slope* represents the K_d. (**c**) Scatter plot representing the diffusion constants versus their fractions from FCCS measurements. *Black symbols* represent the average diffusion constants of the fast and slow components and error bars represent mean ± SD ($n = 50$). The *open upright triangles* (*red*) and *inverted triangles* (*blue*) represent the fast and slow components, respectively

of the K_d value select a cell line which does not endogenously express the target protein, because endogenously expressed protein affects the K_d values, competing with fluorescent labelled proteins. However, the concentration of endogenous proteins in the cells is difficult to determine.

6. Before starting the FCCS analysis, observe the expression of the target protein and its intracellular localization using a fluorescence microscope. For example, p50 and p65 proteins localize in the cytoplasm and also in the nucleus because of their nuclear translocation signal. Set the measuring position in the nucleus or cytoplasm according to interest.

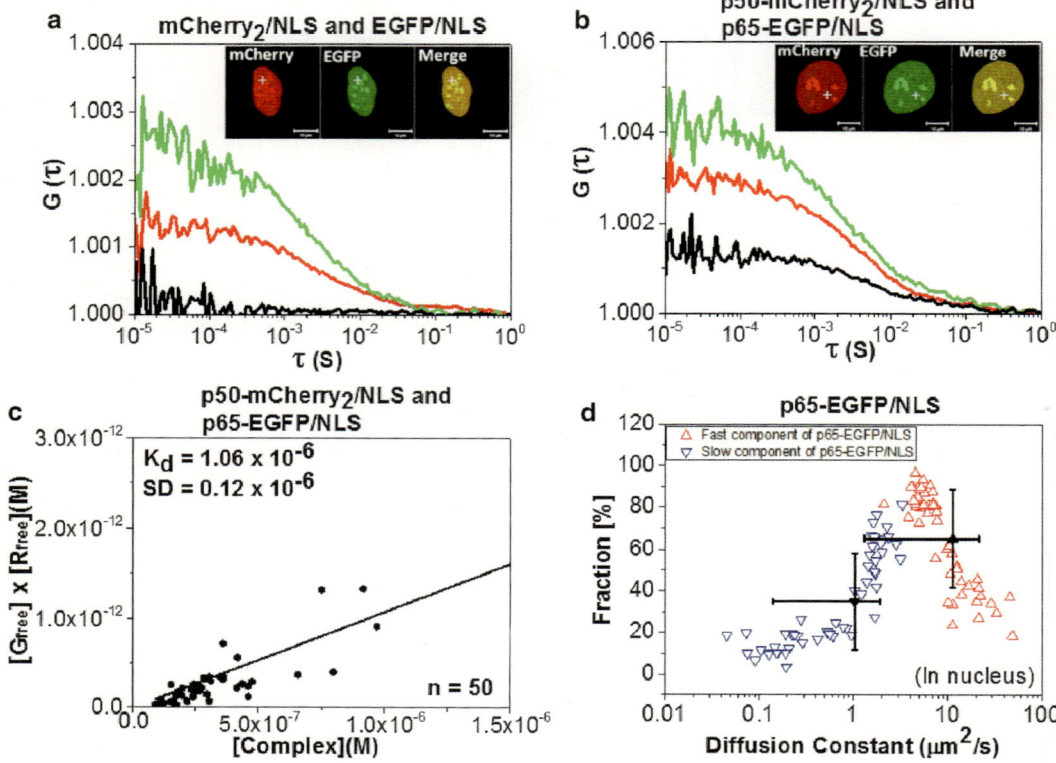

Fig. 7 FCCS measurement and K_d analysis of differently-labeled p50 and p65 in nuclei of living cells. Typical auto- and cross-correlation curves were obtained from cells coexpressing a pair of chimeric fusion proteins. The *green, red* and *black curves* denote the autocorrelation of the green channel ($G_G(\tau)$), autocorrelation of the red channel ($G_R(\tau)$) and the cross-correlation curve ($G_C(\tau)$), respectively. The *insets* show LSM images of the cells coexpressing the pairs of fusion proteins with the measurement positions of FCCS indicated by the white crosshairs. The scale bar represents 10 μm. FCCS measurement are shown of cells coexpressing (**a**) mCherry₂/NLS and EGFP/NLS as a negative control, or (**b**) p50-mCherry₂/NLS and p65-EGFP/NLS. (**c**) results of K_d determination using a scatter plot and linear regression. The plot represents the concentration of free mCherry₂ and EGFP fusion proteins versus the concentration of the complex of mCherry₂ and EGFP fusion proteins. The *solid line* shows a linear fit. The slope represents the K_d. (**d**) Scatter plot representing the diffusion constants versus their fractions from FCCS measurements. *Black symbols* represent the average diffusion constants of the fast and slow components and error bars represent mean ± SD ($n=50$). The *upright open triangles* (*red*) and the *inverted triangles* (*blue*) represent the fast and slow components, respectively

      7. Details of the setup and operation of the ConfoCor 3 system can be found in the ConfoCor 3 Operating Manual (www.zeiss.de/lsm). First adjust the green channel pinhole using only the short-wavelength excitation and the green fluorescent dye standard. Next, adjust the red channel pinhole, using the long-wavelength excitation and the red dye standard. The optimal pinhole diameter depends on the numerical aperture (NA) of the objective lens and the wavelength of each excitation laser. The pinhole size typically selected is equivalent to 1 Airy unit.

8. Select cells that show a low fluorescence intensity to obtain a lower count rate. Optimal intensity of fluorescence and count rates above ~700–800 kHz should be avoided. Very low count rates should be avoided if the background is significant [14].

9. Selection of measurement position is important. Select the position either in the nucleus or the cytoplasm or both. However, different position can be selected in live cells.

10. Lower laser power is recommended for live cell measurements to avoid significant photobleaching of fluorescent proteins or dyes.

11. The structure parameter (s) representing the ratio of the beam waist (2) and the axial radius (1) has a rugby ball shape. These parameters were determined for each green and red channel by measuring Rhodamine 6G and Alexa 594. However, the confocal volume of cross-correlation cannot be determine due to incomplete overlapping of the between green and red channels, which leads to reduced cross-correlation. Thus, the structure parameter for cross-correlation should be fixed to 5.

12. The exact determination of K_d depends on accurate determination of the concentrations of the interacting molecules and thus on the amplitude of the measured auto-correlation function and cross-correlation function. The backgrounds of the resulting number of particles affect the concentration of the interacting molecules.

13. Diffusion time is an indicator of molecular interactions or binding in live cells. The diffusion constant, as a theoretical physical parameter, is derived from the measured diffusion times. Diffusion times are frequently used as parameters of molecular size and as indicators of molecular interactions because they are proportional both to molecular size and to the degree of molecular interactions [13].

14. If the GFP (green fluorescent protein) fusion protein interacts with the RFP (red fluorescent protein) fusion protein, a linear relationship will be obtained when plotting $[G_{free}] \times [R_{free}]$ versus [Complex]. If there is no interaction between the two fluorescent fusion proteins, there will be no linear relationship between $[G_{free}] \times [R_{free}]$ versus [Complex] [5].

References

1. Phelps CB, Sengchanthalangsy LL, Huxford T, Ghosh G (2000) Mechanism of I kappa B alpha binding to NF-κB dimers. J Biol Chem 275: 29840–29846

2. Chen FE, Kempiak S, Huang DB, Phelps C, Ghosh G (1999) Construction, expression, purification and functional analysis of recombinant NFκB p50/p65 heterodimer. Protein Eng 12:423–428

3. Tiwari M, Mikuni S, Muto H, Kinjo M (2013) Determination of dissociation constant of the NFκB p50/p65 heterodimer using fluorescence cross-correlation spectroscopy in the living cell. Biochem Biophys Res Commun 436:430–435

4. Liu P, Ahmed S, Wohland T (2008) The F-techniques: advances in receptor protein studies. Trends Endocrinol Metab 19:181–190

5. Sudhaharan T, Liu P, Foo YH, Bu WY, Lim KB, Wohland T, Ahmed S (2009) Determination of in vivo dissociation constant, K_D, of Cdc42-effector complexes in live mammalian cells using single wavelength fluorescence cross-correlation spectroscopy. J Biol Chem 284:21100

6. Shi XK, Foo YH, Sudhaharan T, Chong SW, Korzh V, Ahmed S, Wohland T (2009) Determination of dissociation constants in living zebrafish embryos with single wavelength fluorescence cross-correlation spectroscopy. Biophys J 97:678–686

7. Liu P, Sudhaharan T, Koh RML, Hwang LC, Ahmed S, Maruyama IN, Wohland T (2007) Investigation of the dimerization of proteins from the epidermal growth factor receptor family by single wavelength fluorescence cross-correlation spectroscopy. Biophys J 93:684–698

8. Savatier J, Jalaguier S, Ferguson ML, Cavailles V, Royer CA (2010) Estrogen receptor interactions and dynamics monitored in live cells by fluorescence cross-correlation spectroscopy. Biochemistry 49:772–781

9. Foo YH, Naredi-Rainer N, Lamb DC, Ahmed S, Wohland T (2012) Factors affecting the quantification of biomolecular interactions by fluorescence cross-correlation spectroscopy. Biophys J 102:1174–1183

10. Oyama R, Takashima H, Yonezawa M, Doi N, Miyamoto-Sato E, Kinjo M, Yanagawa H (2006) Protein-protein interaction analysis by C-terminally specific fluorescence labeling and fluorescence cross-correlation spectroscopy. Nucleic Acids Res 34:e102

11. Glauner H, Ruttekolk IR, Hansen K, Steemers B, Chung YD, Becker F, Hannus S, Brock R (2010) Simultaneous detection of intracellular target and off-target binding of small molecule cancer drugs at nanomolar concentrations. Br J Pharmacol 160:958–970

12. Sadamoto H, Muto H (2013) Fluorescence cross-correlation spectroscopy (FCCS) to observe dimerization of transcription factors in living cells. Methods Mol Biol 977:229–241

13. M.Kinjo, H.Sakata, S. Mikuni, First Step for fluorescence correlation spectroscopy for the living cell. In "Live Cell Imaging -A Laboratory Manual"- Eds Goldman R, Swedlow J, Spector D. Cold Spring Harbor Laboratory Press 2010: 229-238.

14. Bacia K, Schwille P (2007) Practical guidelines for dual-color fluorescence cross-correlation spectroscopy. Nat Protoc 11:2842–2856

15. Muller CB, Loman A, Pacheco V, Koberling F, Willbold D, Richtering W, Enderlein J (2008) Precise measurement of diffusion by multi-color dual-focus fluorescence correlation spectroscopy. Europhys Lett 83:46001

16. Saito K, Ito E, Takakuwa Y, Tamura M, Kinjo M (2003) In situ observation of mobility and anchoring of PKCβI in plasma membrane. FEBS Lett 541:126–131

Chapter 15

Imaging and Quantification of Amyloid Fibrillation in the Cell Nucleus

Florian Arnhold, Andrea Scharf, and Anna von Mikecz

Abstract

Xenobiotics, as well as intrinsic processes such as cellular aging, contribute to an environment that constantly challenges nuclear organization and function. While it becomes increasingly clear that proteasome-dependent proteolysis is a major player, the topology and molecular mechanisms of nuclear protein homeostasis remain largely unknown. We have shown previously that (1) proteasome-dependent protein degradation is organized in focal microenvironments throughout the nucleoplasm and (2) heavy metals as well as nanoparticles induce nuclear protein fibrillation with amyloid characteristics. Here, we describe methods to characterize the landscape of intranuclear amyloid on the global and local level in different systems such as cultures of mammalian cells and the soil nematode *Caenorhabditis elegans*. Application of discrete mathematics to imaging data is introduced as a tool to develop pattern recognition of intracellular protein fibrillation. Since stepwise fibrillation of otherwise soluble proteins to insoluble amyloid-like protein aggregates is a hallmark of neurodegenerative protein-misfolding disorders including Alzheimer's disease, CAG repeat diseases, and the prion encephalopathies, investigation of intracellular amyloid may likewise aid to a better understanding of the pathomechanisms involved. We consider aggregate profiling as an important experimental approach to determine if nuclear amyloid has toxic or protective roles in various disease processes.

Key words Nucleus, Protein aggregation, Fibrillation, Amyloid, Polyglutamine (polyQ) expansion diseases, Neurodegeneration, Huntington's disease, Aging, Congo red

1 Introduction

The term amyloid is of Greek origin and in biomedicine defines abnormally structured proteins that are detectable by specific dyes, antibodies, and a variety of microscopic as well as spectroscopic methods [1, 2]. According to current knowledge, formation of amyloid occurs in a stepwise manner via local unfolding of normally soluble proteins and amyloidogenic intermediates [3–5].

Fibrillation of proteins to amyloid is a hallmark of organismal aging processes [6] and occurs in protein misfolding diseases that

Ronald Hancock (ed.), *The Nucleus*, Methods in Molecular Biology, vol. 1228,
DOI 10.1007/978-1-4939-1680-1_15, © Springer Science+Business Media New York 2015

cover neurodegenerative disorders such as Alzheimer's disease (AD), Parkinson's disease (PD), Huntington's disease (HD), and Creutzfeld–Jakob disease or localized and systemic amyloidoses [7]. While protein fibrillation may appear at the cell surface, e.g., amyloid-β aggregation in AD, or in the cytoplasm, e.g., formation of Lewy bodies in PD, nuclear amyloid is diagnostic of at least nine CAG expansion diseases [8]. Here, proteins such as huntingtin, ataxins 1, 2, 3, and 7, the general transcription factor TATA-binding protein, or the androgen receptor contain unstable CAG repeats of various length that together with heat shock proteins, components of the ubiquitin–proteasome system, and other nuclear proteins are sequestered into insoluble, amyloid-like aggregates in the nucleoplasm [3, 9]. The co-aggregation of endogenous proteins that are involved in protein homeostasis raises the question of the role of such intranuclear amyloid-like inclusions. While clearly representing a diagnostic feature of CAG expansion disorders, it is still unknown whether inclusion bodies and other visible aggregates in the nucleus are pathogenic or protective in neurodegenerative diseases. This is especially true since occurrence of amyloid-like inclusions is not necessarily correlated with neural cell death or functional defects of the "host" neurons [10–13].

We have shown previously that monodisperse silica nanoparticles seed amyloid-like inclusions in the nucleoplasm which are composed of endogenous CAG repeat/polyglutamine (polyQ) proteins, PML nuclear bodies, and components of the ubiquitin–proteasome system [9]. Detection of in situ proteolysis of fluorogenic substrates revealed that at least one-third of these protein aggregates are centers of active proteasomal protein degradation, consistent with a protective role of amyloid-like inclusions by maintaining nuclear protein homeostasis. The importance of such housekeeping becomes even more evident with the demonstration of amyloid structures as intrinsic features of nuclei, i.e., identification of microdomains in the nucleoplasm and in nucleoli that react with amyloid-specific antibodies and dyes [14, 15].

The key to understanding the role of amyloid fibrillation in nuclear function as well as pathology is detection and characterization of its global and local protein composition. The protocols in this chapter address this task, and may aid to correlate features of protein aggregates such as locally active protein degradation with neural fitness and survival in a variety of health or disease settings. Altogether, methods that serve the definition of nuclear amyloid landscapes may identify protective vs. toxic characteristics of protein aggregates.

2   Materials

2.1   Nuclear Amyloid in Cultured Mammalian Cells

2.1.1   Cell Culture

1. Cells: HEp-2 (human epithelioma type 2) and SH-SY5Y (neuroblastoma), purchased from the American Tissue Culture Collection (ATCC).

2. Supplement complete (SC) for HEp-2 cell culture medium: 100 mL of 0.2 M L-glutamine (100×, Gibco), 20 mL of penicillin/streptomycin (100×, Gibco), 100 mL of Na pyruvate (Gibco), 100 mL of nonessential amino acids (MEM 100×, Gibco), 100 mL of RPMI 1640, and 50 µL of 2-mercaptoethanol. Prepare 21 mL aliquots and store at −20 °C.

3. HEp-2 cell culture medium: RPMI 1640 medium supplemented with 10 % fetal calf serum (FCS) and 21 mL of supplement complete (SC).

4. SH-SY5Y proliferation medium: D-MEM/F-12 (1/1) with GlutaMAX (Life Technologies) supplemented with 15 % FCS and 1× penicillin/streptomycin.

5. SH-SY5Y differentiation medium: neurobasal medium (Life Technologies) with B-27 neural cell supplement, 10 µM retinoic acid, 1 % (w/v) L-glutamine, and 1× penicillin/streptomycin.

6. Trypsin–EDTA solution: 0.05 % (w/v) trypsin and 1 mM EDTA.4Na.

2.1.2   Cell Culture Buffers and Solutions

1. Phosphate buffered saline (PBS): 137 mM NaCl, 2.7 mM KCl, 4.3 mM $Na_2HPO_4 \cdot 2H_2O$, 1.4 mM KH_2PO_4.

2. Poly-L-lysine solution: 0.5 mg/mL of poly-L-lysine (mol wt ≥ 300,000) in sterile distilled H_2O.

3. Retinoic acid solution: 20 mM in DMSO.

4. $HgCl_2$ stock solution: 2 mM in sterile PBS (see **Note 1**).

5. Congo red stain: Accustain amyloid stain, Congo red kit (Sigma-Aldrich).

6. Thioflavin T stain: 5 µM Thioflavin T in PBS [16].

7. Lysis buffer: 50 mM Tris–HCl, pH 8.8, 100 mM NaCl, 5 mM $MgCl_2$, 1 mM EDTA, 0.5 % (v/v) Igepal (NP40), and 3 µL/mL of protease inhibitor cocktail (2 mM 4-(2-aminoethyl) benzenesulfonyl fluoride hydrochloride (AEBSF), 0.3 µM aprotinin, 130 µM bestatin, 1 mM EDTA, 14 µM E-64, and 1 µM leupeptin (Sigma-Aldrich)).

8. DNase buffer: 20 mM Tris–HCl. pH 8, 15 mM $MgCl_2$, 0.5 mg/mL DNase I.

9. SDS stock solution: 10 % (w/v) in distilled H_2O.

10. Termination buffer (2×): 40 mM EDTA, 4 % SDS, 100 mM dithiothreitol (DTT).

11. 6 M guanidinium–HCl (Gua-HCl) solution in distilled H_2O.

2.1.3 Cell Culture Equipment

1. Cell culture incubator with humidified atmosphere (37 °C, 5 % CO_2).

2. Glass Coplin jars.

3. Glass coverslips (22 × 22 mm).

4. Tissue culture flasks (250 and 600 mL) and macroplates (6-well).

5. Refrigerated centrifuge: Multifuge (Heraeus) or similar.

6. Refrigerated microcentrifuge: Eppendorf 5417R or similar.

7. Cellulose acetate membranes: pore size 0.2 μm (Whatman).

8. Dot-blotting device (Bio-Dot, Bio-Rad).

9. Microscope: laser scanning confocal system with 60×/1.4NA Plan Apo objective (Olympus IX70 Fluoview or similar system).

10. Image analysis software: MetaMorph 4.6 (Molecular Devices, Sunnyvale, CAL, USA).

2.2 Nuclear Amyloid in Caenorhabditis elegans (C. elegans)

2.2.1 Strains

1. N2 wild type *C. elegans* (provided by the Caenorhabditis Genetics Center (CGC), which is funded by the NIH Office of Research Infrastructure Programs (P40 OD010440)).

2. *Escherichia coli (E. coli)* strain OP50.

2.2.2 Reagents, Buffers, and Solutions

1. Nematode Growth Medium (NGM) plates: add 750 mL of distilled H_2O to 15 g of BD Bacto Agar, 2.25 g of NaCl, and 1.9 g of BD Bacto Proteose No. 3. Supplement with 3.75 g of BD Bacto yeast extract. Autoclave to sterilize, cool down to 55 °C, and add 750 μL of solution A, 375 μL of solution B, 750 μL of solution C, and 18.75 mL of solution D. Mix carefully by shaking the flask and pour the NGM into 9 cm-diameter petri dishes under sterile conditions. After drying overnight, spread 500 μL of an overnight culture of *E. coli* strain OP50 in the center of the plate with an inoculation spreader (*see* **Note 2**).

2. Solution A: dissolve 0.5 g of cholesterol in 100 mL of EtOH.

3. Solution B: dissolve 11.08 g of $CaCl_2$ in 100 mL of distilled H_2O, autoclave.

4. Solution C: dissolve 24.65 g of $MgSO_4 \cdot 7H_2O$ in 100 mL of distilled H_2O, autoclave.

5. Solution D: 108.3 g of KH_2PO_4, 36 g of K_2HPO_4 in 1 L of distilled H_2O, autoclave.

6. 5-Fluoro-2′-deoxyuridine (FUDR) solution: 50 mM in distilled water, store at −20 °C. To keep the worms culture age-synchronized, add FUDR to a final concentration of 40 µM to cold (<55 °C) NGM prior to pouring the plates.

7. Worm buffer M9: 3 g of KH_2PO_4, 6 g of Na_2HPO_4, 0.5 g of NaCl, and 1 g of NH_4Cl in 1 L of distilled H_2O. Sterilize by autoclaving.

8. Synchronization solution: 2 mL of 4 M NaOH, 3 mL of 12 % (w/v) NaClO (*see* **Note 3**), and 5 mL of distilled H_2O.

9. Silica nanoparticles (silica NPs): 50 nm-diameter (Kisker, Steinfurt, Germany).

10. Formaldehyde solution: 10 % ultrapure methanol-free EM grade (Polysciences, Warrington, PA, USA, *see* **Note 1**).

11. Liquid nitrogen (*see* **Note 4**).

12. Acetone (synthesis grade) and methanol (extra pure, *see* **Note 1**).

13. Phosphate buffered saline (PBS) (*see* Subheading 2.1.2, **item 1**).

14. Congo red solution: dissolve Congo red (Sigma) in PBS at 7 mg/mL, filter through a 0.2 µm syringe filter to remove insoluble aggregates (*see* **Note 9**).

15. VECTASHIELD mounting medium (Vector Laboratories).

16. Nail polish.

2.2.3 Equipment

1. Dissecting microscope.

2. Worm picker: flame the tip of a glass Pasteur pipette and melt a 32-G platinum wire into the glass [17].

3. Petri plates: 9 cm-diameter with ventilation cams.

4. Pasteur pipettes, scalpel, staining jars.

5. Incubator at 20 °C to culture *C. elegans*.

6. Cytoslides (Shandon); glass coverslips 22 × 22 mm, 0.15 mm thick.

7. Fluorescence microscope.

3 Methods

3.1 Nuclear Amyloid in Cultured Mammalian Cells

3.1.1 Amyloid Staining

Visualization and quantification of amyloid and amyloid-like protein aggregates in the cell nucleus can be realized with different labeling methods (Fig. 1, left) (*see* **Note 5**). The first option is to use amyloid-specific dyes such as Congo red (CR) or Thioflavin T (ThT). Both these dyes are well established and widely used for clinical diagnostics of amyloid in neurodegenerative protein misfolding diseases. Because of this major application these dyes are used mostly for histological staining, but they are likewise applicable for cell culture. Additionally, development of antibodies has enabled

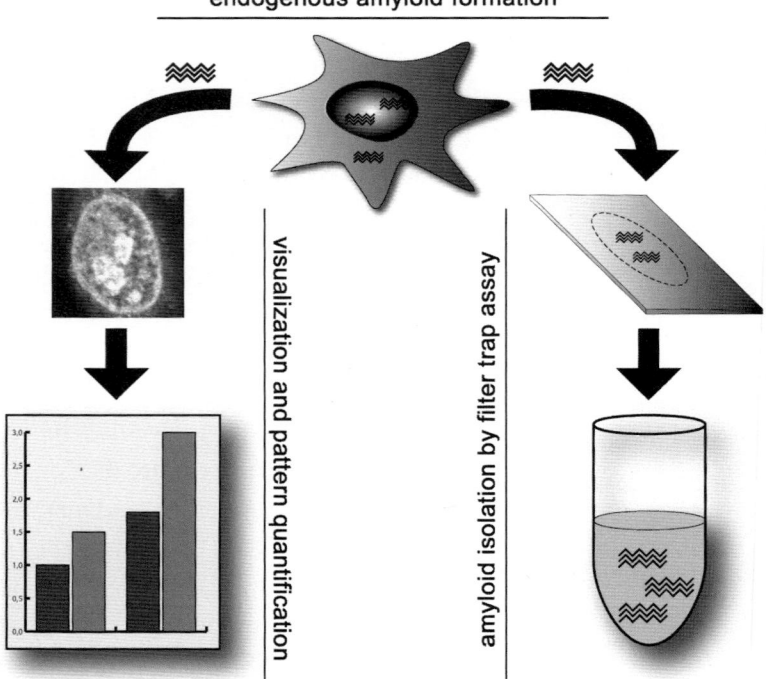

Fig. 1 Schematic workflow for analyzing nuclear amyloid in cultured mammalian cells. Fibrillar aggregation of endogenous proteins can be induced by subtoxic concentrations of $HgCl_2$ [14]. *Left*, amyloid-specific dyes (Congo red, Thioflavin T) or the antibody WO1 that specifically detects amyloid structures [1] enable localisation of amyloid in cultured cells. *Right*, for biochemical analysis of amyloid, SDS-insoluble protein aggregates are isolated by filter retardation assays and dissolved in guanidinium-HCl (Gua-HCl)

specific labeling of amyloid structures by indirect immunofluorescence. The mouse monoclonal IgM antibody WO1 was established to detect amyloid forms of amyloid-β [1]. As this antibody is not sequence-specific but structure-specific, it recognizes amyloid structures formed by other proteins as well. For both dye- and antibody-based detection methods, the cells have to be fixed on glass coverslips and permeabilized to allow dyes or antibodies to enter the cell and nucleus. Primary antibodies (WO1) are detected by secondary antibodies conjugated to a fluorophore. Different fluorophores are commercially available and should be chosen with respect to experimental settings and microscope/detection setup.

Culture of Mammalian Cells

1. Culture HEp-2 cells in RPMI 1640 medium and SH-SY5Y cells in D-MEM/F-12 (1/1) medium to 80 % confluence (37°C, 5 % CO_2, humidified atmosphere) (*see* **Note 6**).

2. Harvest the cells by trypsination, spin down ($300 \times g$, 4 °C, 8 min), and determine the cell number.

3. SH-SY5Y and other neuronal cell lines require coverslips coated with poly-L-lysine to support adherence and differentiation, but HEp-2 cells can be cultured on uncoated coverslips. To coat coverslips with poly-L-lysine, wash them in 100 % EtOH and sterilize by flaming. Overlay them with poly-L-lysine solution and incubate at room temperature for 10 min. Remove the solution, wash the coverslips 3× by dipping into sterile distilled H_2O, and let them dry completely.

4. Place a coverslip in each well of 6-well plates and seed 1.5×10^5 cells per well.

5. HEp-2 cells can be used for experiments after ~48 h, when they should be at 80 % confluence. SH-SY5Y cells differentiate to post-mitotic neuronal phenotypes when transferred to serum-free Neurobasal medium containing retinoic acid (10 μM) (*see* Subheading 2.1.1, **item 5**). Culture SH-SY5Y cells in proliferation medium for 24 h, change the medium to differentiation medium, and culture for an additional 5–7 days in this medium with changes every 48 h (*see* **Note 7**).

6. On the day of the experiment add $HgCl_2$ to the culture medium at a mild, subtoxic concentration (50–60 μM for HEp-2 cells, 20–25 μM for SH-SY5Y cells) for 4 h to induce aggregation of endogenous nuclear proteins, and leave control cells untreated (*see* **Note 8**).

Staining with Congo Red or ThT

1. Transfer coverslips into a glass Coplin jar and wash them with PBS.

2. Fix the cells with methanol (at –20 °C) for 5 min.

3. Permeabilize the cells with acetone (at –20 °C) for 2 min. Avoid drying of the cells.

4. Wash with PBS (3 × 10 min).

5. Apply the Congo red staining kit protocol according to the manufacturer's instructions (*see* **Note 9**), or incubate with ThT staining solution for 10–20 min.

6. Wash with PBS (3 × 10 min).

7. Place each coverslip cell-side down on a drop of mounting medium on a slide and seal the edges with nail polish. Store the slides at 4 °C in the dark.

Immunolabeling with Antibody W01

1. Follow **steps 1–4** of Subheading "Staining with Congo Red or ThT".

2. Incubate the cells with primary antibody (1:50 in PBS) for 1 h in a humidified chamber.

3. Transfer the coverslips into a glass Coplin jar and wash with PBS (3 × 10 min).

4. Incubate the coverslips with secondary, fluorophore-conjugated antibodies (diluted in PBS) for 45 min in a humidified chamber.

5. Transfer the coverslips into a glass Coplin jar and wash with PBS (3×10 min).

8. Place each coverslip cell-side down on a drop of mounting medium on a slide and seal the edges with nail polish. Store the slides at 4 °C in the dark.

Microscopy

1. Cells stained for nuclear amyloid can be imaged by confocal microscopy (*see* **Note 10**).

2. Save the image data as tiff files without data compression.

Quantification of Amyloid Fluorescence Patterns

1. Confocal images can be analyzed using the linescan tool of MetaMorph 4.6 or comparable software. Scan lines of interest (LOIs) at defined positions through the cell nucleus and save the respective intensity values in log files.

2. Import the linescan data to Microsoft Excel. The data has to contain the pixel number of the line p_n and the corresponding fluorescence intensity values x_n.

3. Interpret the line data as a graph $G = (V, E)$ according to the graph theory of discrete mathematics. Vertices v are defined by measured points (p_n, x_n) and edges e by the distance between two measured points. Use the following equations:

$$e = \{v_n, v_{n+1}\} \tag{1}$$

$$e_n = \sqrt{(p_n - p_{n+1})^2 + (x_n - x_{n+1})^2} \tag{2}$$

4. To calculate a pattern describing the parameters, define the range of interest r which defines the graph vertices and edges used:

$$r(v_a; v_b) = p_b - p_a + 1 \tag{3}$$

To characterize the complete nuclear pattern, use $a = n_{min} = 1$ and $b = n_{max}$. If you want to analyze only nuclear regions containing aggregates, a and b are defined by intensity thresholds to detect and analyze only high intensity objects.

5. Calculate absolute values of intensity and distance within the defined range:

$$i(r(v_a; v_b)) = \sum_{n=a}^{b} x_n \tag{4}$$

$$d(r(v_a; v_b)) = \sum_{n=a}^{b} e_n = \sum_{n=a}^{b} \sqrt{(p_n - p_{n+1})^2 + (x_n - x_{n+1})^2} \tag{5}$$

6. With the calculated values for range r, intensity I, and distance d it is possible to calculate values for the average fluorescence intensity I and the average fluorescence heterogeneity H of the nuclear amyloid pattern:

$$I = \frac{i\left(r\left(v_a; v_b\right)\right)}{r\left(v_a; v_b\right)} \tag{6}$$

$$H = \frac{d\left(r\left(v_a; v_b\right)\right)}{r\left(v_a; v_b\right)} \tag{7}$$

7. Use values (I and H) calculated for each single cell to compare changes of nuclear amyloid patterns in different experimental groups. Perform calculations for at least 100 cells per group to enable statistical analysis.

8. For further description of these analysis methods and results, *see* [14].

3.1.2 Isolation of Amyloid by Filter Retardation Assays

As well as visualizing them, the characterization of nuclear protein aggregates requires their isolation and purification (Fig. 1, right). Amyloid-like protein aggregates represent large, SDS-insoluble macromolecular assemblies that are retrievable via filter trapping. Using a filter retardation assay [18], cell lysates are diluted in SDS and filtered through cellulose acetate membranes with a pore size of 0.2 μm. These membranes have a low protein-binding capacity and trap only large SDS-insoluble aggregates while soluble cell components pass the pores, ensuring aggregate retention by size only, in contrast to nitrocellulose membranes which have high protein-binding properties. With this method the trapped aggregates can be isolated and used for further experiments.

1. Culture cells according to Subheading "Culture of Mammalian Cells", but use 600 mL cell culture flasks instead of 6-well plates.

2. Wash untreated and $HgCl_2$-treated cells very rapidly with PBS. Long incubation with PBS may lead to cell detachment and loss, especially when using SH-SY5Y cells.

3. Detach the cells with 2 mL of trypsin–EDTA solution at 37 °C and harvest them in 40 mL of PBS at 4 °C.

4. Spin the cells down by centrifugation (300×g, 4 °C, 8 min) and discard the PBS.

5. Resuspend the cells in 10 mL of PBS, determine the cell number, and repeat **step 4**.

6. Use $6 \times 10^6 - 1.2 \times 10^7$ cells per experiment. Add 1–1.5 mL of lysis buffer to the cell pellet and resuspend by pipetting up and down.

7. Incubate for 30 min on ice (*see* **Note 11**).

8. Isolate the insoluble cell fraction by centrifugation in a microfuge ($15,000 \times g$, 5 min, 4 °C).

9. Discard the supernatant, dissolve the pellet in 200 μL of DNase buffer, and incubate for 1 h at 37 °C.

10. Add 200 μL of termination buffer (2×) and heat to 98 °C for 5 min.

11. Dilute aliquots of the samples further in 2 % SDS in order to titrate the optimal amyloid concentration for membrane filtering. If the concentration of SDS-insoluble aggregates is too high, overloading the membrane filter could impair filtration and reduce specificity.

12. Filter the samples (200 μL/well) through a cellulose acetate membrane (pore size 0.2 μm) on a Bio-Dot device by applying a vacuum (*see* **Notes 12** and **13**).

13. Wash the membrane twice with 0.1 % SDS (200 μL/well).

14. Remove the membrane and air-dry. Use immediately or store at –20 °C.

15. Cut out the dots with trapped protein aggregates from the membrane, place them in 2.0 mL Eppendorf tubes, and add 100–150 μL of 6 M Gua-HCl to each tube (*see* **Note 14**). Incubate the tubes overnight on a shaker.

16. Transfer the eluate into a new 1.5 mL tube and use the isolated amyloid aggregates for further analyses, for example by mass spectrometry.

3.2 Nuclear Amyloid in C. elegans

3.2.1 C. elegans Culture

These transparent nematodes are maintained on Petri dishes of NGM medium that are seeded with a lawn of OP50, a uracil-auxotroph *E. coli* strain [17, 19]. A prerequisite for most experiments is to keep the worms age-synchronized, and this is achieved via isolation of the eggs using bleach and culturing the worms from the L4 larval stage on FUDR-supplemented plates. FUDR inhibits DNA synthesis and prevents self-fertilization [20, 21].

1. Use a worm picker to transfer 8–12 adult *C. elegans* onto six fresh 9 cm-diameter NGM plates supplemented with yeast extract, and culture at 20 °C for ~3 days until the plates contain ~500 gravid hermaphrodites (*see* **Note 15**).

2. Harvest the worms by washing them off the plates with buffer M9, collect them in a 15 mL tube by centrifugation at $1,000 \times g$ for 30 s, and discard the supernatant.

3. Add 5 mL of synchronization solution (*see* **Note 3**) to the worm pellet, resuspend the pellet by inverting the tube 3× (*see* **Note 16**), spin down at $3,000 \times g$ for 30 s, and discard the supernatant (*see* **Note 17**).

4. Repeat **step 3**.

5. Wash the worm pellet 3× with distilled H_2O, and spin down at $3,000 \times g$ for 30 s.

6. Discard the supernatant after the last washing step and distribute the egg pellet onto six new 9 cm NGM plates supplemented with yeast extract and seeded with OP50.

7. Incubate the eggs at 20 °C. They develop into L4 larvae after ~44 h [22] (*see* **Note 18**).

8. Wash the L4 larvae off the plates with M9 and transfer them into a 15 mL tube. Let the larvae sink to the bottom and remove the buffer with a pipette.

9. Transfer the pellet to six fresh 9 cm NGM plates supplemented with yeast extract and 40 µM FUDR and seeded with OP50, and incubate the worms for ~15 h at 20 °C.

3.2.2 Feeding C. elegans with Silica NPs

Silica NPs induce amyloid-like protein aggregates in the nucleoli of intestinal cells of *C. elegans* [20]. One-day old, age-synchronized, adult *C. elegans* are placed onto silica NP-supplemented NGM plates and incubated at 20 °C. After three days, amyloid-like protein aggregates are detectable via staining with the amyloid-specific dye Congo red (Fig. 2) (*see* Subheading 3.1.1).

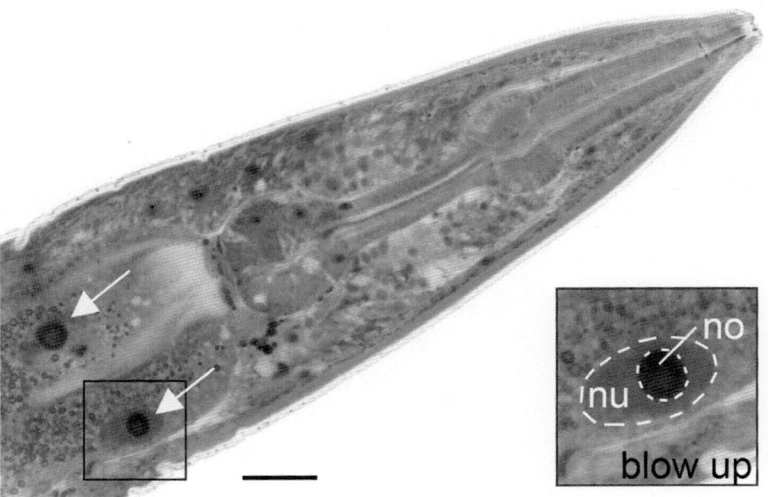

Fig. 2 Nuclear amyloid in *C. elegans*. Pharynx and anterior intestine of a representative silica NP-treated, 4-day-old wild type worm. Amyloid-like protein aggregates are visualized in the fixed worm via Congo red staining. *Arrows* indicate silica NP-induced amyloid-like protein aggregates in nucleoli of anterior-most intestinal cells. The blowup shows one representative nucleus of an intestinal cell. *No* nucleolus, *nu* nucleus. Scale bar 20 µm

1. Seed 9 cm-diameter NGM plates containing FUDR with 500 μL of OP50 supplemented with 50 nm-diameter silica NPs (2.5 mg/mL), or with the same volume of distilled H_2O for controls.

2. Incubate the plates overnight at 37 °C and cool down to 20 °C.

3. Collect 1-day old, adult *C. elegans* by wash-off with M9, transfer them with a 1.5 mL Pasteur pipette into a 15 mL tube, and let them sink to the bottom. Remove the buffer with a pipette.

4. Wash the worms 3× with M9 to separate and discard younger worms and already-laid eggs. Let the worms sink to the bottom and remove the buffer.

5. Transfer the worms onto two NGM plates, one supplemented with silica NPs and one with distilled H_2O.

6. Incubate at 20 °C and transfer the worms every day to new NGM plates with silica NPs or distilled H_2O.

3.2.3 Fixation of C. elegans

In order to prepare the worms for staining with the amyloid-specific dye Congo red, 4-day old adult hermaphrodites are fixed via a freeze-crack method to conserve the tissue and to enable the penetration of the stain [23].

1. Coat the ~12 mm diameter sample chambers (only) of cytoslides with poly-L-lysine (*see* Subheading "Culture of Mammalian Cells", **step 3**); do not flame these slides. Wash 1× with distilled H_2O after the EtOH wash and let the cytoslides dry.

2. Wash H_2O- or silica NP-treated 4-day old worms off with M9, collect them in a 1.5 mL tube, and let them sink to the bottom.

3. Remove OP50 by washing the worms 3× with M9. After the last wash, discard the supernatant and leave the worms in ~100 μL of M9.

4. Mix 5 μL of freshly prepared 8 % formaldehyde in PBS with 5 μL of worm suspension in the poly-L-lysine-coated sample chamber of a cytoslide, using the pipette tip (*see* **Notes 1** and **19**).

5. Place a coverslip onto the worms and press it gently with forceps to immobilize them on the cytoslide (*see* **Note 20**).

6. Fix the worms for 5 min and transfer the slide with forceps into a bowl with liquid nitrogen to shock-freeze them for some minutes (*see* **Note 4**).

7. Remove the slides sequentially from the liquid nitrogen, carefully push a scalpel blade under one side of the coverslip to loosen it, and detach it with a fast movement (*see* **Note 21**).

8. Refix the worms by transferring the slide to a staining jar containing methanol at −20 °C (*see* **Note 1**). Incubate for 5 min in a freezer at −20 °C.

9. Place the slides into a new jar filled with acetone at –20 °C. Incubate for 2 min in a freezer to permeabilize the worms (*see* **Notes 1** and **22**).

10. With forceps, transfer the slides successively into three new jars filled with PBS to wash off remaining methanol and acetone and to rehydrate the worms.

3.2.4 Congo Red Staining

1. Place one slide into a humidified chamber after draining excess PBS with a tissue.

2. Add 150 μL of Congo red solution and proceed with the remaining slides in the same way.

3. Incubate the slides in the humidified chamber for 1–2 h at room temperature.

4. Remove the Congo red solution with a pipette and wash the worms 4× in PBS (*see* **Notes 22** and **23**).

5. Drain excess PBS with a tissue, place two drops of VECTASHIELD over the worms, and cover them carefully with a coverslip (*see* **Note 24**). Cover the slides with a tissue and apply soft pressure to drain excess VECTASHIELD.

6. Seal the coverslip with nail polish and let dry completely. Slides can be stored at 4 °C in the dark.

7. Detect nuclear amyloid in the intestinal cells by fluorescence microscopy. Congo red can be excited with 568 nm light and fluoresces at 585–640 nm. A representative micrograph of Congo red-stained 4-day old *C. elegans* is shown in Fig. 2 (*see* **Note 25**).

4 Notes

1. HgCl₂, methanol, acetone, and formaldehyde are toxic and have to be handled with care and according to institutional safety regulations.

2. Never damage the surface of the NGM plate or spread *E. coli* to its edges, to prevent *C. elegans* from crawling into the agar or off the plate, respectively.

3. Be aware that NaOH and NaClO are corrosive.

4. Working with liquid nitrogen requires appropriate protection (thick gloves, safety goggles, lab coat).

5. Please note that the three amyloid labeling methods presented, Congo red staining, ThT staining, and immunodetection with the antibody WO1 show slightly different specificity in HgCl₂-treated cells. While Congo red labels larger areas of the nucleoplasm and already shows a weak homogenous nucleoplasmic staining in untreated cells, labeling with WO1 antibody only detects

more distinct regions of the nucleoplasm. Therefore Congo red detects amyloid aggregates as well as intermediates, oligomers, or other amyloid-like structures. In contrast, the WO1 antibody specifically detects fibrillated protein aggregates. Thus, the appropriate labeling method has to be chosen with respect to the aim of the study.

6. The relatively large nuclei of HEp-2 cells are advantageous to investigate nuclear subdomains and to apply quantitative fluorescence analysis. HEp-2 cells were used to establish the methods for Congo red pattern quantification [14], which can also be applied to RA-differentiated SH-SY5Y cells.

7. There are several protocols to induce neuronal differentiation of SH-SY5Y cells. As retinoic acid supplementation is the best-characterized and most widely used, it is described here. Alternative protocols can be found in ref. 24.

8. The protocol presented here uses $HgCl_2$ at a mild concentration to induce nuclear protein aggregation in HEp-2 or SH-SY5Y cells. You have to verify the subtoxicity of the $HgCl_2$ treatment for other cell types by performing viability assays.

9. Test different concentrations of Congo red (100, 50, 20, or 10 %) to optimize the staining. The intensity and specificity of staining can be slightly modulated by ethanol washing steps (*see* the staining kit protocol). Mild washing results in higher fluorescence intensity but may lower specificity due to higher background staining, whereas extended washing may reduce the fluorescence and impede proper pattern quantification.

10. If you wish to quantify fluorescence patterns, especially their heterogeneity, it is necessary to use confocal or similar microscopy techniques. Otherwise, out-of-focus signals will interfere with intensity measurements of nuclear subdomains and therefore mask changes of fluorescence pattern.

11. The aggregate isolation protocol presented here includes pre-fractionation of buffer-insoluble components to promote proper filtration. Depending on the experimental setup and task it is possible to use other cell fractions, for example whole cell extracts or nuclear fractions. To isolate nuclear fractions, replace Subheading 3.1.2, **step 7** by incubation for 10 min on ice, when only the cell membrane is lysed and cytoplasmic components detach (vortex and pipet up and down to aid detachment). Subsequent centrifugation will isolate cell nuclei in the pellet.

12. Problems may occur during filter retardation assays if large DNA or DNA fragments are still intact. In general, you will recognize the presence of large DNA since it is precipitated by SDS after addition of termination buffer. The pores of the membrane may be blocked, flow of the liquid will stop or slow

down, and non-amyloid components will be trapped on the membrane. If this problem occurs, increase the volume of DNase buffer (and correspondingly of termination buffer), but note that this also dilutes the isolated protein aggregates. An alternative to DNA digestion is sonication of the sample before addition of termination buffer.

13. Sometimes samples may filter very fast when applying the vacuum, due to low protein concentration or leakage between the membrane and the Bio-Dot device.

14. Amyloid protein aggregates dissolve in 6 M Gua-HCl as well as in 70 % formic acid. If you want to elute aggregates from membranes as described in this protocol, use of formic acid is not recommended because cellulose acetate filters are not resistant.

15. Avoid starving adult *C. elegans* hermaphrodites, because starved worms develop an internal hatch phenotype and eggs cannot be isolated anymore. Food should always be abundant.

16. Tap the tube with your fingers to loosen the pellet.

17. To control the isolation procedure, check whether the hermaphrodites have released their eggs. Be careful, because a too long incubation in synchronization solution will harm the eggs. Proceed when the first eggs appear outside the hermaphrodites.

18. After ~44 h the eggs develop into L4 larvae, distinguishable by a characteristic white triangle at the developing vulva. It is important to monitor the development in every experiment.

19. Five microliter of worm suspension should contain ~20–30 worms. Pipetting up and down allows adjustment of the number of worms in the sample volume.

20. It is necessary to control the pressure because during the staining procedure the worms should neither burst, nor get lost due to low adhesion.

21. Handle with care and protect your hands from burning by liquid nitrogen.

22. Never change the solutions in the staining jars during fixation, permeabilization or washing steps, because the worms detach easily. Always transfer the slides with forceps from one solution to the next.

23. It is possible to further stain the DNA of the worms with DAPI: add 4 μL of DAPI solution (5 mg/mL) to 100 mL of PBS and use this solution in the last washing step. Incubate for 10 min.

24. Air bubbles should be avoided.

25. Be aware of autofluorescence and always include an unstained negative control. Only analyze worms with good preservation of the tissue.

References

1. O'Nuallain B, Wetzel R (2002) Conformational Abs recognizing a generic amyloid fibril epitope. Proc Natl Acad Sci U S A 99:1485–1490

2. Uversky V, Lyubchenko Y (2014) Bionanoimaging: protein misfolding and aggregation. Elsevier, San Diego

3. Ross CA, Poirier MA (2005) Opinion: what is the role of protein aggregation in neurodegeneration? Nat Rev Mol Cell Biol 6:891–898

4. Chiti F, Dobson CM (2009) Amyloid formation by globular proteins under native conditions. Nat Chem Biol 5:15–22

5. Volpatti LR, Vendruscolo M, Dobson CM et al (2013) A clear view of polymorphism, twist, and chirality in amyloid fibril formation. ACS Nano 7:10443–10448

6. David DC, Ollikainen N, Trinidad JC et al (2010) Widespread protein aggregation as an inherent part of aging in C. elegans. PLoS Biol 8:e1000450

7. Blancas-Mejía LM, Ramirez-Alvarado M (2013) Systemic amyloidoses. Annu Rev Biochem 82:745–774

8. Nelson DL, Orr HT, Warren ST (2013) The unstable repeats-three evolving faces of neurological disease. Neuron 77:825–843

9. Chen M, Singer L, Scharf A et al (2008) Nuclear polyglutamine-containing protein aggregates as active proteolytic centers. J Cell Biol 180:697–704

10. Saudou F, Finkbeiner S, Devys D et al (1998) Huntingtin acts in the nucleus to induce apoptosis but death does not correlate with the formation of intranuclear inclusions. Cell 95:55–66

11. Gutekunst CA, Li SH, Yi H et al (1999) Nuclear and neuropil aggregates in Huntington's disease: relationship to neuropathology. J Neurosci 19:2522–2534

12. Arrasate M, Mitra S, Schweitzer ES et al (2004) Inclusion body formation reduces levels of mutant huntingtin and the risk of neuronal death. Nature 431:805–810

13. Tsvetkov AS, Ando DM, Finkbeiner S (2013) Longitudinal imaging and analysis of neurons expressing polyglutamine-expanded proteins. Methods Mol Biol 1017:1–20

14. Arnhold F, von Mikecz A (2011) Quantitative feature extraction reveals the status quo of protein fibrillation in the cell nucleus. Integr Biol 3:761–769

15. Arnhold F, von Mikecz A (2014) Intranuclear amyloid - local and quantitative analysis of protein fibrillation in the cell nucleus. In: Uversky V, Lyubchenko Y (eds) Bionanoimaging: protein misfolding and aggregation. Elsevier, San Diego, pp 475–484

16. LeVine H III (1993) Thioflavine T interaction with synthetic Alzheimer's disease β-amyloid peptides: detection of amyloid aggregation in solution. Protein Sci 2:404–410

17. Stiernagel T (1999) Maintenance of *C. elegans*. In: Hope IA (ed) *C. elegans*. A practical approach. Oxford University Press, Oxford, NY, pp 51–67

18. Wanker EE, Scherzinger E, Heiser V et al (1999) Membrane filter assay for detection of amyloid-like polyglutamine-containing protein aggregates. Methods Enzymol 309:375–386

19. Brenner S (1974) The genetics of *Caenorhabditis elegans*. Genetics 77:71–94

20. Scharf A, Piechulek A, von Mikecz A (2013) The effect of nanoparticles on the biochemical and behavioral aging phenotype of the nematode *Caenorhabditis elegans*. ACS Nano 7:10695–10703

21. Honda Y, Tanaka M, Honda S (2010) Trehalose extends longevity in the nematode *Caenorhabditis elegans*. Aging Cell 9:558–569

22. Altun ZF, Hall DH (2009) Introduction. In: WormAtlas doi:10.3908/wormatlas.1.1

23. von Mikecz A, Scharf A (2013) Isochronal visualization of transcription and proteasomal proteolysis in cell culture or in the model organism *Caenorhabditis elegans*. Methods Mol Biol 1042:257–273

24. Kovalevich J, Langford D (2013) Considerations for the use of SH-SY5Y neuroblastoma cells in neurobiology. Methods Mol Biol 1078:9–21

Analysis of Nuclear Organization with TANGO, Software for High-Throughput Quantitative Analysis of 3D Fluorescence Microscopy Images

Jean Ollion, Julien Cochennec, François Loll, Christophe Escudé, and Thomas Boudier

Abstract

The cell nucleus is a highly organized cellular organelle that contains the genome. An important step to understand the relationships between genome positioning and genome functions is to extract quantitative data from three-dimensional (3D) fluorescence imaging. However, such approaches are limited by the requirement for processing and analyzing large sets of images. Here we present a practical approach using TANGO (Tools for Analysis of Nuclear Genome Organization), an image analysis tool dedicated to the study of nuclear architecture. TANGO is a generic tool able to process large sets of images, allowing quantitative study of nuclear organization. In this chapter a practical description of the software is drawn in order to give an overview of its different concepts and functionalities. This description is illustrated with a precise example that can be performed step-by-step on experimental data provided on the website http://biophysique.mnhn.fr/tango/HomePage.

Key words Nucleus, Genome, 3D fluorescence microscopy, Quantitative image analysis software, Statistical analysis

1 Introduction

The genome is highly organized within the nucleus. Numerous studies have shown that spatial positioning of the genome within the nuclear space plays an important role in major cellular functions, such as regulation of gene expression, DNA replication and repair [1]. Two complementary families of approaches are used to study 3D organization of the genome: molecular assays (such as 3C-derivates or ChIP-derivates) and microscopy-related techniques. The latter have the advantage of providing direct spatial information at the single-cell level. Fluorescence microscopy techniques such as immunocytochemistry and Fluorescence In Situ Hybridization (FISH) allow selective

Ronald Hancock (ed.), *The Nucleus*, Methods in Molecular Biology, vol. 1228,
DOI 10.1007/978-1-4939-1680-1_16, © Springer Science+Business Media New York 2015

fluorescent labeling of multiple compartments and DNA sequences within the nucleus.

Though recent microscopes facilitate the acquisition of large sets of 3D images, the automatic processing and analysis of these images remains a difficult task that often requires programming skills. To overcome this limitation, several software programs have been developed. In this paper the focus is on TANGO, a software for high-throughput quantitative analysis of 3D fluorescence microscopy images and dedicated to the study of nuclear organization [2]. It provides functionalities for the different steps of the process of analysis of fluorescences images: image processing, quantitative measurements, and statistical analysis of measurement results. Since it is dedicated to the analysis of nuclear organization, its usage could be made easy (no programing skills are required) while maintaining a high level of versatility within this field of research. This versatility is due to the possibility of adapting processing to different types of signals, to perform many quantitative measurements, and to the fact that it is integrated into two generic environments: ImageJ for image processing and quantitative measurements and R for statistical analysis of measurement results. Its potential is also increased by its modularity that allows it to be enhanced by plug-ins. An intuitive graphical user interface allows one to rapidly verify the quality of segmentation of images, reducing the time devoted to this step which is usually the most time-consuming.

In this chapter the TANGO software is described first in order to give an overview of its concepts and functionalities. In the second part, the focus is on a precise example that can be performed step-by-step on experimental data provided on the website http://biophysique.mnhn.fr/tango/MIMB.

2 Overview of the Software

2.1 General Description

2.1.1 Structure of the Software

TANGO is composed of three elements (Fig. 1):

- An ImageJ plug-in for image processing and quantitative measurements.

- A MongoDB database for storing settings, images, selection of subpopulations (*see* Subheading "Sub-populations"), and measurement results.

- An R package for statistical analysis of measurement results.

ImageJ and R are both very modular, offer numerous functionalities, and are widely used in the scientific community. Therefore, when using TANGO one benefits directly from the rich potential of those two environments.

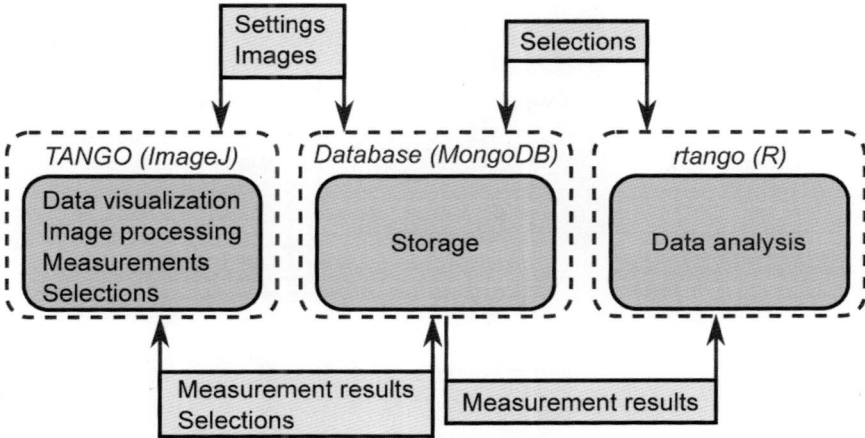

Fig. 1 Structure of the TANGO software. Each element of the software and its environment is represented in a *dotted rectangle* and its function in the *green rectangle*. The type of data exchange between environments is represented in the *yellow rectangles*, and the direction is indicated by the *arrows* (Color figure online)

2.1.2 Image Processing and Analysis Work Flow

As TANGO is dedicated to the analysis of nuclear architecture, the global work-flow is constrained (Fig. 2): first nuclei are segmented from a DNA-stained image (typically stained with DAPI or Hoechst), then all fluorescence channels are cropped around each segmented nucleus. Then, objects contained in subnuclear structures can be segmented. This work-flow yields a hierarchical data structure reflecting the physical objects: an *experiment* contains *fields*, that contain *nuclei*, that contain *structures* that contain segmented *objects*. The same data structure is used in the database, in the graphical user interface (*see* Subheading "Accessing the Different Level of Objects"), and in the R package (*see* Subheading 2.3).

Segmentation is achieved by applying a highly customizable processing chain (*see* Subheading 2.2.2) to an image, which allows it to adapt to different labeled structures. After image processing, measurements are performed at the level of individual nuclei (*see* Subheading 2.2.4).

2.1.3 Documentation

We provide several sources of documentation:

- Tutorials on the website http://biophysique.mnhn.fr/tango/HomePage.

- Documentation on processing and analysis modules included in TANGO on the website http://biophysique.mnhn.fr/tango/HomePage.

- Internal documentation that can be accessed through the *Help* button in the connect tab (this needs to be downloaded from the website by clicking on the *Update* button).

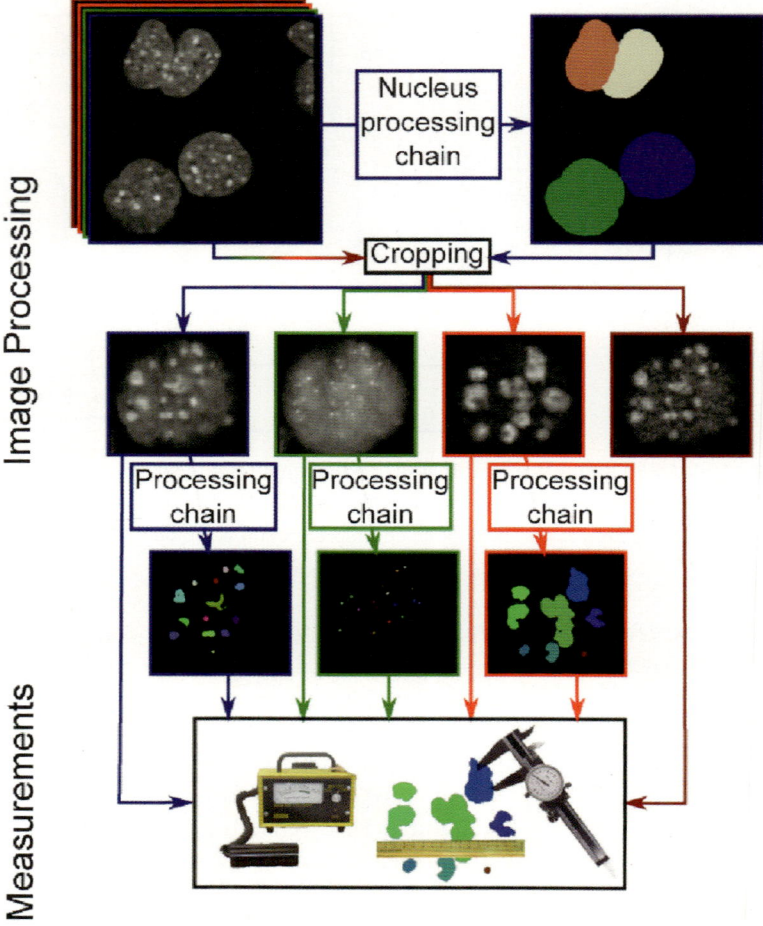

Fig. 2 Image processing workflow of TANGO

2.2 Image Processing and Analysis

In this section we describe in detail how to perform image processing and measurements on images. A plug-in system allows one to add processing operations and measurements to TANGO. *See* the website http://biophysique.mnhn.fr/tango/DevManual for details on plug-in development.

2.2.1 Experiment

An experiment corresponds to a set of images with the same set of signals. Setting up an experiment is achieved in the *Edit Experiment* tab. First, the number of fluorescence channels is set (further termed channel images), then the structures are defined, each of them being associated to a channel image (*see* **Note 1**). The first structure is always the nucleus and the others are subnuclear structures. All further processing and analysis is performed on structures. In the example we provide, there are four channel images corresponding to four fluorescence channels (DAPI, FITC,

Cy3, and Cy5) and four structures (nucleus, Ki67, centromeres, and nucleoli). By default, each structure is associated with the channel of the same index, but it can be changed when editing structures. Once the structure of the experiment is defined, images can be imported. All formats included in bio-formats (*see* **Note 2**) are supported. All imported images must have exactly the same number of channels in the same order as defined in the experiment.

<table>
<tr><td>*2.2.2 Processing Chain*</td><td>A processing chain (PC) designates the sequence of processing</td></tr>
</table>

A processing chain (PC) designates the sequence of processing operations that are performed on a raw image in order to segment objects. They can be used to segment nuclei or to segment sub-nuclear structures. This is generally the limiting step in the whole analysis process, and defining a PC requires knowledge of image processing [2]. In TANGO, a processing chain is composed of three successive steps: *pre-processing*, *segmentation*, and *post-processing*, that corresponds to three categories of processing operations (Table 1). Each step can be customized by adding processing operations of the corresponding category (Fig. 3).

- Pre-processing usually aims at correcting common imaging artifacts, reducing noise, and enhancing the signal-to-noise ratio.

- Segmentation is the process of identifying objects in the pre-processed image. Several algorithms for segmentation of nuclei and of nuclear signals have been implemented in TANGO, such as the seed-based 3D watershed algorithm.

- Post-processing allows for correction to the segmented image. Classical operations are removing of small objects and morphological operations such as closing.

In order to automate processing over large sets of nuclei, some processing operations can be parametrized with automatic thresholding methods. This way, the value of a threshold can be adjusted to each image instead of choosing a constant value. This increases the robustness of processing over field-to-field or cell-to-cell intensity variations, and allows one to use the same parametrization for different experiments.

PCs are set up in the *Edit Processing Chains* tab. The conception of a PC is usually done through a sequence of trial/error. In order to simplify this process, each step can be easily tested on images thanks to the *Test* button located on the right of each processing operation (*see* **Note 3**). Because of the image size difference, the behavior of the test is different for nuclei processing and structure processing. For nuclei processing, only the selected step will be applied on the active image. For structure processing, the selected step and all previous steps are applied on the selected nucleus (visible in the *Data* tab). When a parameter of a processing

Table 1
Main processing operations included in TANGO

Operation	Type	Usage/properties
Median 3D	Pre-processing	De-noising (edge-preserving)
Gaussian 3D	Pre-processing	De-noising
Laplacian of Gaussian 3D	Pre-processing	Spot enhancement
Subtract Gaussian	Pre-processing	Background reduction (useful to correct for inhomogeneous illumination)
Top-hat 3D	Pre-processing	Spot enhancement
Hysteresis segmentation	Segmentation	Two-threshold-based segmentation
Simple segmentation	Segmentation	Threshold-based segmentation
Nucleus edge detector	Segmentation	Nucleus segmentation based on edge detection (robust over intensity variations)
Seeded watershed 3D	Segmentation	Watershed 3D seeded with regional minima
Spot detector 3D	Segmentation	Seeded watershed 3D based on Hessian transform (for segmentation of small spots)
Spot segmenter	Segmentation	Segment spots using seeds, watershed, and local thresholding
Morphological filters 3D	Post-processing	3D morphological filters performed individually on objects (opening, closing, 2D/3D fill holes) optimized for speed for large radii
Size filter	Post-processing	Erase objects on the basis of a size criterion

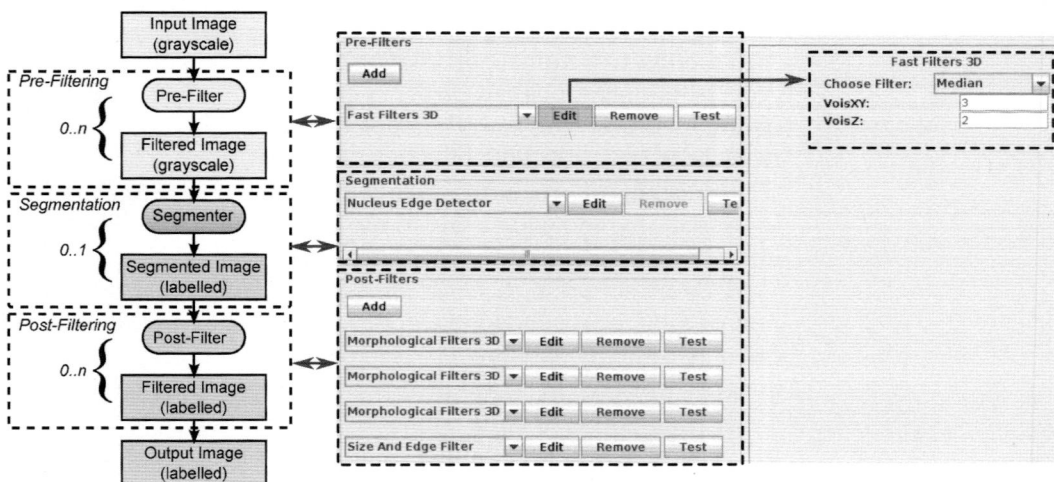

Fig. 3 Processing chains. A processing chain is composed of three steps, *pre-processing*, *segmentation*, and *post-processing* (*see* main text). The *right panel* shows the processing chain editing interface. For each of the three steps, processing operation can be chosen among operations of the corresponding category. If needed, parameters can be adjusted (*rightmost arrow*)

operation is not correctly defined, it appears in red. Templates can be defined, saved, and assigned to any structure of any experiment of the current user. Differences between a processing chain and its template are shown in blue. PCs templates can be imported and exported, using the buttons located in the *Connect* tab.

2.2.3 Semiautomatic Processing of Nuclei

Automatic processing of nuclei often fails to correctly separate nuclei close to one another, such as in dense tissue sections. A semiautomatic procedure was designed for 3D segmentation of nuclei that takes approximately 1 min per nucleus for an experienced user. The details of the procedure are described in a tutorial on our website (*see* **Note 4**).

The principle of the procedure is the following: the first step is to isolate the nucleus from others. For that purpose, the user draws a selection on several planes around the nucleus. The drawing must be precise at the contact sections of neighboring nuclei, but can be rough elsewhere. Contours of upper, mid and lower planes (and any other planes) are drawn manually, and intermediate planes are automatically interpolated, thus defining a 3D mask that contains the nucleus. A more precise adjustment of this mask to the nucleus can be done by performing a processing chain within the mask.

2.2.4 Measurements

Measurements allow for quantification of various features of a signal (Table 2). They can be classified in two categories:

- Segmentation-based measurements that operate on segmented objects. These measurements include shape descriptors of objects (volume, surface, elongation, etc.), and descriptors of spatial relations between objects such as distance, overlap, etc.

- Voxel-based measurements, which do not rely on segmentation and allow one to quantify various aspects of raw signals, such as the moments of the histogram, texture parameters, radial-autocorrelation, or voxel-based colocalization between two signals.

Measurements are set up in the *Edit Experiment* tab, *Measurement* sub-tab (*see* Fig. 4). Each measurement can be tested on a selected cell thanks to the *Test* button. Measurements are performed from the *Data* tab, either from the *Field* frame on the nuclei contained in the selected fields or in the *Cell* frame on the selected cells. If the *Override* check-box is not selected, only new measurements will be performed.

2.2.5 Data Browsing

Accessing the Different Level of Objects

Figure 5 shows the lists that allow one to access the different hierarchical levels: fields, cells, structures and segmented objects.

- The *Fields* frame (Fig. 5a) allows one to access the different microscopy fields contained in an experiment, perform operations on fields (crop, process images, perform measurement), and display the whole field images.

Table 2
Main measurement modules included in TANGO

Name of module	Short description	Reference
Distances	Center–center, center–border, border–border distance between objects	[3]
Grayscale spatial moments	3D spatial moments of signal	[3]
JACOP colocalization	Voxel-based colocalization	[4]
Measure geometrical, advanced	Measurements of 3D surfaces of segmented objects using meshes	[3]
Measure geometrical, simple	Geometrical measurements on segmented objects (volume, elongation, feret, etc.)	[3]
Eroded volume fraction	Normalized distance measurement to nuclear periphery	[5]
Signal quantification	Statistical descriptors of signal intensity within objects (mean value, min, max, quantile, etc.)	[3]
Texture 3D	Classical matrix texture descriptor (entropy etc.)	[3]
Radial-autocorrelation 3D	3D radial-autocorrelation of signal	[3]
Spatial analysis	Spatial statistics tools adapted to the study of nuclear organization, allowing one to test the randomness of a spatial distribution of points	[6]

- The *Cells* frame (Fig. 5b) can be displayed from the *Fields* frame by clicking on the *Cells* button. It allows one to access the segmented nuclei within the selected fields, perform operations (process images, perform measurements) on cells and display cropped images or 3D representations of segmented objects. Cells are represented by a thumbnail that enables rapid inspection of the shape of the nuclei (or other structures). Cells can be annotated with a tag that will change its color in the list. Selections (*see* below) are an alternative way for annotations.

- The *Objects 3D* frame (Fig. 5c) can be displayed from the *Cells* frame by clicking on the *Objects* button. It permits access to the segmented objects within the selected structures, and to perform basic operations on segmented objects (delete, merge, split). When clicking on an object in the list, its contours are displayed on the active image and updated when changing the slice index (Fig. 5d). This functionality is very useful to rapidly verify segmentation.

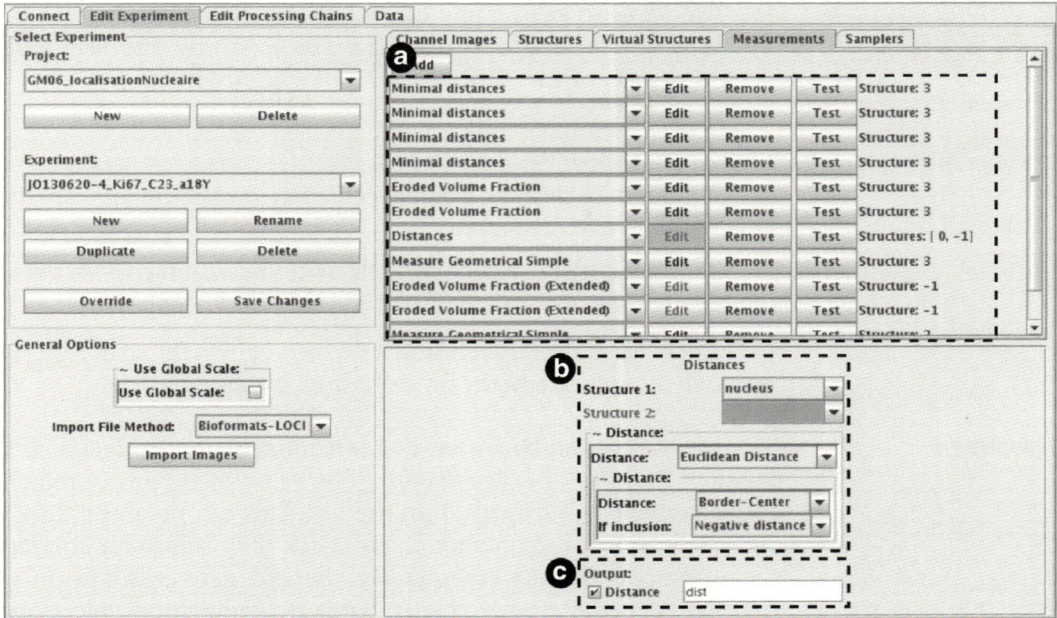

Fig. 4 Measurements setup. (**a**) List of measurements to be performed on the cells of the current experiment. Parameters of each measurement can be edited from the *Edit* button. (**b**) Parameter edition interface: in this example, the measurement computes distances between objects of two structures. The parameters are the two structures and the type of distance (center-to-center, border-to-border, etc.). Incorrectly set parameters appear in red (here the second structure is missing). The measurements return one output value (in this case an array of distances between all the objects of each structure), whose name can be edited (**c**)

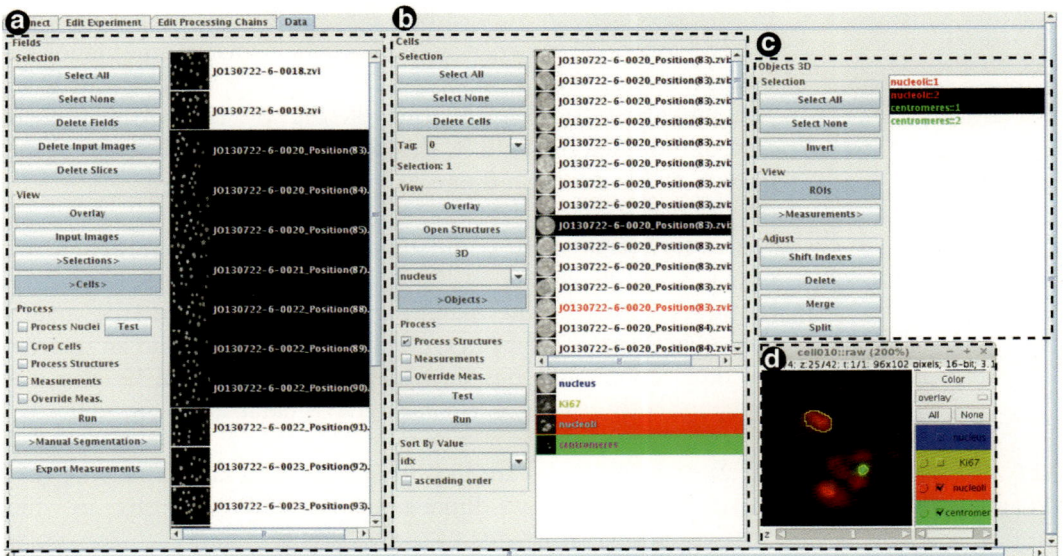

Fig. 5 Data browsing interface. (**a**) Field frame. (**b**) Cell frame; the *upper list* contains cells, the *lower list* contains structures of the selected cell. (**c**) Object frame. (**d**) Example of image with contours of selected objects displayed in yellow

- The *Measurement* frame can be displayed from the *Object 3D* frame by clicking on the *Measurements* button. It authorizes the display of the different types of measurements:
  - When no structures/objects are selected, all measurements are displayed.
  - When a single object is selected, measurements on this object are displayed.
  - When two objects are selected, measurements between pairs of objects are displayed (e.g., distances).

When no object is selected but one or several structure are selected, all measurements on structures are displayed.

Sorting Lists

The list of cells can be sorted by measurements on nuclei (such as volume, number of objects, etc.). The value of the measurement for each nucleus is displayed on the list. This can be useful during the verification of segmentation to check only cells with aberrant measurements (in our example, we checked only cells with more than two centromeres detected). Similarly, segmented objects can be sorted by measurements on objects.

Subpopulations

Subpopulations of nuclei or objects (termed *selections*) can be defined in the "selection" frame (displayed by clicking on the *Selections* button of the *Cells* frame). When clicking on a selection, the selected nuclei/objects are highlighted in gray in the corresponding lists. The *Show Only* button will hide all cells not included in the selected selection. Nuclei or objects can be added or removed from a selection. In Subheading 3.3.2 selections will be seen that are the means of a two-way link between images in ImageJ and measurement results in R because they can be imported in R and generated from both R and ImageJ (Fig. 6). If a selection is created or modified through R, changes can be updated by the *Update* button of the selection frame.

2.3 Statistical Processing with R

In this section the basic commands to manage measurement results under R will be explained. Basic knowledge of the R language is required, especially the manipulation of dataframes (the main R data structure). R has been embedded in TANGO and an interface for basic commands has been designed, but R can be also used independently as long as the mongoDB database is running.

2.3.1 Importing Measurement Results in R

Measurement results are imported into R as dataframes. A dataframe is a table, or two-dimensional array-like structure, in which each column corresponds to a measurement and each row corresponds to one case. There are two commands allowing to import different types of measurements. They generate a dataframe that contains one or several measurements of the same type and performed on the same structure(s):

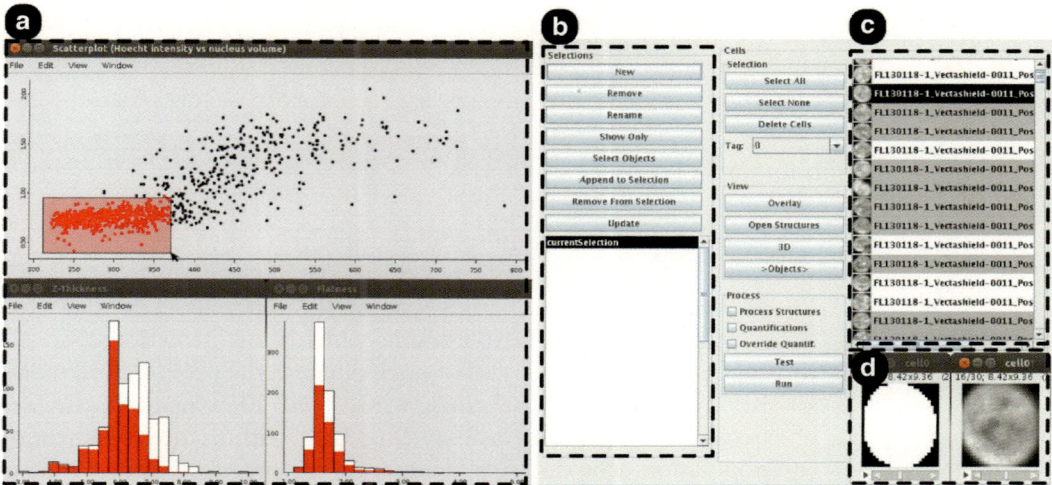

Fig. 6 Selection of subpopulations in ImageJ and R. (**a**) Features from nuclei measured in TANGO can be imported into R using the r-package rtango, and plotted using the r-package IPlots (http://www.rosuda.org/iplots/, package for interactive plots). In this case, each point of the scatter plot represents one nucleus. When a subpopulation is selected, it is interactively updated on other plots. (**b**) Selections can be exported from R to TANGO, nuclei contained in the selection will be *highlighted* in the nucleus list (**c**) and the associated images can be visualized (**d**) or processed separately. Reciprocally, selections created or edited in TANGO are accessible within R, allowing one to compare different subpopulations

- *extractObjectData*: generates a dataframe of measurements performed on objects of a given structure that yield one single value per object (e.g., the volume of the object). Two indexes are necessary to locate a value, the index of the nucleus and the index of the object within the nucleus.

- *extractStructureData*: generates a dataframe of measurements performed on one or on several structures. In the case of a measurement between pairs of objects (e.g., distance between two objects), three indexes are required to locate a value: the index of the nucleus and the indexes of each object. This command can also extract measurements with a single value per nucleus, or any number of values.

2.3.2 Managing Subpopulations

Subpopulations of nuclei or objects (termed *selections*) can be defined within TANGO (*see* Subheading "Sub-populations") and imported into R as a list of object indexes. This permits annotation of a dataframe, and thus comparison of the distribution of measurements from different subpopulations.

Selections of nuclei/objects can also be generated within R and saved in the database so that they will be further accessible through the graphical interface of TANGO, or within R. This can be carried out through a specific command, and alternatively through the use of interactive plots (Fig. 6). This permits display of the images corresponding to a given range of measurement values and thus can be very useful to study outliers, for instance.

3 Methods: Image Analysis Protocol

Here we provide a biological example related to the study of centromere localization with respect to the nuclear border and nucleoli in a population of fixed lymphoblastoid cells. The data come from immuno-FISH experiments which allow detection of specific centromeres by FISH and of nucleolin and the proliferation marker Ki67 by immunocytochemistry. Ki67 will allow us to discriminate between cycling and non-cycling cells within the asynchronous population. The distance measurement is an adaptation of the normalized distance termed the eroded volume fraction (EVF) [3]. The adaptation implemented in TANGO makes it possible to take into account the spatial extension of the centromere, and to measure distances from other structures than the nuclear periphery (the nucleoli in our case) [7].

Subheading 3.1 is dedicated to the installation and configuration of TANGO. To illustrate the image processing and analysis capabilities of TANGO, an example is detailed in Subheading 2.2 with a sample image containing several nuclei and adapted processing chains provided. Subheading 3.3 illustrates the statistical analysis of the example with a whole population of processed nuclei provided. Online material can be found on the website http:// biophysique.mnhn.fr/tango/MIMB.

3.1 Installation and Startup

TANGO is available for Linux/Windows/MacOSX 64-bit systems (*see* the website for requirements and details on installation). The latest version of TANGO should be used. As described in Subheading 2.1.1, TANGO is composed of three parts, a plug-in for ImageJ, a MongoDB database, and a package for R. The MongoDB database is an independent software that has to be running in order to access data from ImageJ or R. It can be run from the ImageJ TANGO interface, or installed as a service on the same computer as ImageJ or on a distant server.

To start TANGO, start ImageJ and in the menu *TANGO* select *TANGO > tango*. This will display the main interface of TANGO. At startup, TANGO checks that the installation is complete; if it is not, messages will be displayed in the log. The first step is to connect TANGO to the database: enter the host address of the server in the *Host* text area (leave it blank if the database is running on the same computer) and click on the *Connect* button.

The ImageJ interface of TANGO can be used to start the database server on the local computer: in the *Connect* tab, if this has not already been done, set the directory containing the executables of MongoDB (through the *Set MongoDB Binary Directory* button) and set the directory to which MongoDB should write the data (through the *Set Database Directory* button). Click on the *Start MongoDB* button.

3.2 Image Processing and Analysis

Online supporting material (*see* **Note 5**):

- Sample image.
- Sample processing chains.

1. *Create a user account* (button *Add User* of the *Connect* tab).

2. *Creation of the experiment.* In the *Edit Experiment* tab:

   (a) Create a project and an experiment.

   (b) In the *Channel Images* sub-tab, add four channel files and edit their names (*keyword*) (DAPI, A488, Cy3, Cy5).

   (c) In the *Structures* sub-tab, add four structures and edit their names and associated channel images (first: nucleus/DAPI, second: Ki67/Cy3, third: Nucleoli/Cy5, fourth: Centro7/A488). The order of structures does not necessarily correspond to the order of channel files (*see* Subheading "Image Processing and Quantitative Measurements"). The display color of each structure can be set here.

   (d) Click on the *Save Changes* button.

3. *Import the sample image*:

   (a) Leave the default import image method as Bioformat-LOCI.

   (b) Click on the *Import Images* button and select the downloaded sample image.

4. *Import the PC templates*:

   (a) Download and unzip the processing chains.

   (b) In the *Connect* tab, click on *Import Processing Chains* and select the folder containing the processing chains (.bson files).

5. *Set the processing chains for each structure*: in the *Edit Processing Chains* tab, set the previously imported template for each structure and click on the *Copy from template* button.

   (a) For the structure "Nucleus" set the template "Nucleus". This PC will first apply a median filter to reduce noise (while preserving edges), the segmentation algorithm is based on edge-detection and thus robust to cell-to-cell intensity variations within a field, and post-filters will remove small objects and fill holes (nucleoli).

(b) For "Ki67" no PC is set, because there is no need to detect objects.

(c) For "Nucleoli" use the "Nucleoli" template. This PC will first apply a median filter to reduce noise, and perform a thresholding using Otsu's method [8]. Post filters will remove small objects.

(d) For "centro7" use the "centroFISH" template. The pre-filters will reduce noise and enhance local contrast of spot-like objects. Then a watershed transform is applied on the gradient of the image, and non-spot objects are removed by intensity and Hessian (intensity curvature) criteria. The thresholds of those criteria are computed for each image.

(e) Click on the *Save* button.

6. *Run the processing and default measurements.* In the *Data* tab, *Fields* frame, *Process* sub-frame: select the field to be processed in the field list, check *Process nuclei, Crop cells, Process structures* and *Measurements*. Click on the *Run* button: a progress window will appear while the process is running.

7. *Verification of segmentation.* After processing, different objects can be accessed in the *Data* tab in order to verify the quality of processing (*see* Subheading "Accessing the Different Level of Objects").

(a) To verify the contours of segmented objects, they can be displayed on the image. In the *Cells* frame (accessed through the button *Cells* in the *Field* frame), select a cell, select the structure "Nucleoli", and click on the *Open Structure* button to open the image (or select a previously opened image). In the *Object 3D* frame (accessed through the *Objects* button of the *Cells* frame), click on an object: its contours will be displayed on the selected image and updated when changing slices.

(b) In the example, exactly two centromeric signals are expected per cell. To avoid opening all cells to verify the segmentation, it is possible to check only cells with more or less than two signals. For that purpose, in the *Cells* frame select "objectNumber_3" in the *Sort By Value* sub-frame. The list of cells will be sorted by the number of objects in the third structure (Centro7) and the number of centromeres written after the name of the cells in the list.

(c) When a cell contains a defect (for instance the nucleus is cut, there are too many centromeres, etc) it can be annotated with a tag (select the cell and change the *tag* in the *Cells* frame). Setting the tag to "−1" will exclude the cell from processing and measurements. Alternatively, selections can be used to annotate cells or objects (*see* Subheading "Sub-populations").

8. *Set the measurements.* In the *Measurements* sub-tab of the *Edit Experiment* tab:

   (a) Normalized distance to periphery: add "Eroded Volume Fraction (Extended)", select the *structure* "Centro7", check *erode nuclear edges*. Write "Nucleus_" in the *prefix* text area of the output frame below. This will add "Nucleus_" before the name of the measurement.

   (b) Normalized distance to nucleoli: add "Eroded Volume Fraction (Extended)", select the *structure* "Centro7", select "Nucleoli" as the *reference structure*, and check *negative distance inside internal structures*. Write "Nucleoli_" in the *prefix* text area of the output frame below.

   (c) Average intensity of Ki67 within nucleus: add "Signal Quantification", set *structure object* as "Nucleus", *structure signal* as "Ki67", and output *prefix* as "Ki67_".

   (d) Aggregation of Ki67: add "Radial Autocorrelation 3D", set *structure* as Ki67, *number of filters* as 1, *#1 filter* as "Misc 3D Filters" (leave the default settings: Gaussian 3D with a radius of 2), *number of radius* as 1, *#1 Radius* as 2, for output, set *Autocorrelation radius 1* as "Ki67_aggregation".

   (e) Click on the *Save changes* button.

9. *Run the measurements.* In the *Data* tab, *Fields* frame: select the field to be processed in the field list, check *Measurements*. Click on the *Run* button.

10. *Data export.* Measurements can be accessed from the *Data* tab (*see* Subheading "Accessing the Different Level of Objects"), exported to a spreadsheet (*Export Measurement* button of the *Fields* frame in the *Data* tab), or directly imported from R (*see* below).

3.3 Statistical Analysis with R

Online supporting material:

- Sample nucleus population.
- R script.
- Multiple-experiment R script.

R is a very powerful software for statistical analysis of data. It contains many functionalities for manipulation of large datasets, for graphical representations, and for performing statistical tests. In this section, functionalities that are directly related to TANGO will be presented (they are located into the R-package *rtango*). We recommend Rstudio (https://www.rstudio.com/) as user interface for R, the R package *plyr* (http://plyr.had.co.nz/) for data manipulation (*see* Subheading 3.3.3), and the R package *ggplot2* (http://ggplot2.org/) for graphical representations. Basic commands can be performed from the *Analysis* tab of the user interface. All actions that correspond to R commands are visible in the integrated R console of TANGO, so that they can be copied in a script to automate analysis.

The following sections detail the steps of the analysis. The name of the step is in italics, and the corresponding step from the provided R script is indicated in brackets. Each step of the script contains one or several commands that have to be run the in the R console. Text after the "#" symbol corresponds to comments and will not be executed by the R console.

3.3.1 Peripheral Positioning of a Centromere

In this section we explain how to connect R with TANGO, import data to R, and create a plot.

1. *Import sample data.* Download the sample nucleus population and unzip the folder. In the *Connect* tab, click on the *Import Data* button, select the unzipped folder, and enter a name for the imported project. The import is complete when the interface unfreezes, which might take a few minutes.

2. Connection with the database (step 1.1). In R, edit the R script and modify the following variables with the actual values: *host. name* (if the database is on the same computer as R, leave "localhost"), *folder.name* (the name of the folder containing the experiment), *experiment.name* (the name of the experiment). Execute the commands of step 1.1 in the R console. This will create the session and experiment objects stored in the variable *session* and *experiment*, respectively. Several functions can be called from these objects through the "$" symbol (*see* below).

3. *Data import from the database to R* (step 1.2). The first command creates a list of nuclei for which tags are included in [0; 10] and that contain 1 or 2 centromeres. The second line creates a dataframe containing the measurements from the previously selected nuclei.

4. *Generate a cumulative distribution plot (step 1.3).*

3.3.2 Subpopulations

Here, the previous example is refined by defining two subpopulations corresponding to quiescent and cycling cells, based on analysis of the Ki67 signal [7]. Then there is a comparison of the two subpopulations on a single plot. Steps 2.3–2.5 can be performed for each measurement.

1. Extraction of Ki67 data (step 2.1). This will generate a dataframe containing the two measurements used to analyze the Ki67 signal: "Ki67_aggregation" which is the aggregation of Ki67, and "Ki67_average" which is the average intensity of

Ki67. Both measurements are considered as measurements on nuclei because they are both performed within the nuclear space and yield one single value per nucleus.

2. *Definition of subpopulations.* Subpopulations are recorded in TANGO and accessible in the interface, and thus one must define subpopulations only once. They can be accessed and edited from TANGO (*see* Subheading "Sub-populations"). There are two ways of defining subpopulations:

(a) Graphical method (step 2.2.1). The two first commands of step 2.2.1 generate an interactive plot. Select the quiescent population ("Ki67_aggregation" < *aggregation.threshold*) and run the third command, the selection will be recorded in the database as "G0". Idem for cycling cells: select the subpopulation ("Ki67_average" > *intensity.threshold*) and run the fourth command.

(b) *Command-line method (step 2.2.2).* This method is equivalent, but the subpopulations are explicitly defined with the R command "subset" which authorizes combinations of logical operations and thus a more complex definition of a subpopulation than the previous method.

(c) *Annotation of dataframe (step 2.3).* This will add a column containing the name of the subpopulation ("G0" or "Cycling") for each row, which permits distinguishing them in a plot.

(d) *Generation of a plot (step 2.4).* This will generate a plot in which the two subpopulations are represented by two curves of different colors.

(e) *Comparison of the two subpopulations (step 2.5).* A dataframe is first generated for each subpopulation, and then the two distributions of measurements are compared using the Kolmogorov–Smirnov test.

3.3.3 Advanced Manipulation of Dataframes

In this section, we provide a few hints to manipulate dataframes. The operations presented are not directly related to the *rtango* package, but are very useful when manipulating data generated with TANGO.

Combination of Dataframes

TANGO generates dataframes that contain only measurements of the same type, but it can be useful to combine several dataframes in order to analyze relations between measurements of different types. In the example, a dataframe containing measurements on centromeres will be combined to a dataframe containing measurement on nuclei, in order to analyze the relation between the peripheral localization of centromeres and the nuclear volume.

1. *Merge the two dataframes (step 3.1).* The first dataframe has several rows per nucleus (corresponding to each centromere) and the second one exactly one per nucleus. Thus, when merging the two dataframes, rows of the second one will be replicated. The column shared by the two dataframes is "nucleus.id" and is specified to the merge function. Several columns can be provided to the merge function if needed.

2. *Study the relation between the two variables (step 3.2).* The commands will generate a scatter plot and perform a correlation test, taking into account the cell-cycle or not.

Operations on a Dataframe

R provides very useful functions to perform operations on dataframes. An example is provided of using the function *ddply* of the package *plyr*. This function splits a dataframe, performs a customizable operation on each sub-dataframe, and merges the resulting sub-dataframes into a new dataframe.

For instance, it can be considered that when only one centromere is detected in a nucleus, this means in fact that the two alleles are merged into one spot. To remove the potential bias, the corresponding rows will be duplicated to have exactly two rows per nucleus.

1. *Definition of the operation (step 4.1).* First create a custom function that does the operation per nucleus, and then apply this function on the whole dataframe. In this step, create a function that inputs a dataframe and outputs a dataframe with duplicated rows if the input dataframe has only one row.

2. *Application of the operation (step 4.2).* This will generate a new dataframe with duplicated rows that can be used in steps 2.4 and 2.5 instead of the previous dataframe.

3. *Study of the effect of duplication (step 4.3).* In order to study differences between the distribution with duplicated rows and the unprocessed distribution, a new column is added to the dataframes containing the duplicated and un-duplicated distributions in order to distinguish them and these are concatenated in one dataframe for plotting.

3.3.4 Management of Several Experiments

The dataset provided contains two experiments corresponding to two different centromeres. An R script is provided allowing one to analyze them together in order to compare them. The content of the script is not detailed in this section; only a few hints are provided to manage several experiments at once.

Image Processing and Quantitative Measurements

- The settings of an experiment (channel images, structures, processing chains, measurements) can be duplicated to create a new experiment initialized with the same settings.

- In order to manage different experiments the same way, the structure order should be the same in all experiments. As the structure and channel files are uncoupled, this should not be a constraint.

- If an experiment has been modified (e.g., a measurement is added), the modifications can be updateded on other experiments thanks to the O*verride* command in the *Edit Experiment* tab. Measurements should be performed after modifying the list of measurements of an experiment.

- Processing and measurements can be performed on multiple experiments at once in batch mode using the console of TANGO (*see* http://biophysique.mnhn.fr/tango/BatchProcessing).

Statistical Analysis
- Usage of a script is necessary to automate the analysis. The user interface can be used to help creating the script (as all actions are recorded in the console) but can not be used to directly manage several experiments.

- As shown in the multiple-experiment script, different experiments can be stored in a list containing specific information for each experiment. Then each operation (extracting data, annotating dataframes, etc.) is performed by iterating on the list of the experiment.

- Subpopulations are defined once for each experiment and are stored in the database, and thus can be retrieved from a script.

- Finally, all dataframes are merged in one single dataframe that contains all the data, and test/plot are performed using the merged dataframe.

To summarise, automated analysis of large sets of images usually requires the development of laborious scripts both for image analysis and statistical processing. TANGO integrates powerful software like ImageJ for the image processing aspect, MongoDB for fast and reliable database storage, and the R environment for efficient statistical analysis. This integrated software greatly facilitates medium-content analysis and can be used as a versatile tool for everyday image processing and statistical studies.

4   Notes

1. The main advantage of uncoupling channels and structures is that a single channel can be processed differently in order to detect different objects. For instance, in mouse cells DAPI staining can be used to detect nuclei but also the bright spots termed chromocenters contained within nuclei.

2. http://loci.wisc.edu/software/bio-formats.

3. In test mode, the verbose level is increased and intermediate steps are usually displayed.

4. http://biophysique.mnhn.fr/tango/Manual+Nucleus+Segmentation+Tutorial.

5. If more details are needed, see the tutorial "getting started" on our website http://biophysique.mnhn.fr/tango/HomePage.

Acknowledgements

We thank P. Andrey for fruitful discussions and advice.

References

1. Cavalli G, Misteli T (2013) Functional implications of genome topology. Nat Struct Mol Biol 20:290–299

2. Russ J (2011) The image processing handbook, 6th edn. CRC Press, Boca Raton, FL

3. Ollion J, Cochennec J, Loll F et al (2013) TANGO: a generic tool for high-throughput 3D image analysis for studying nuclear organization. Bioinformatics 29:1840–1841

4. Bolte S, Cordelières FP (2006) A guided tour into subcellular colocalization analysis in light microscopy. J Microsc 224:213–232

5. Ballester M, Kress C, Hue-Beauvais C et al (2008) The nuclear localization of WAP and CSN genes is modified by lactogenic hormones in HC11 cells. J Cell Biochem 105:262–270

6. Andrey P, Kiêu K, Kress C et al (2010) Statistical analysis of 3D images detects regular spatial distributions of centromeres and chromocenters in animal and plant nuclei. PLoS Comput Biol 6:e1000853

7. Ollion J, Loll F, Cochennec J et al (2014) Proliferation dependent positioning of individual centromeres in the interphase nucleus of human lymphoblastoid cell lines. Mol Biol Cell (revision)

8. Otsu N (1979) A threshold selection method from gray-level histograms". IEEE Trans Sys Man Cyber 9:62–66

Chapter 17

Quantitative Analysis of Chromosome Localization in the Nucleus

Sandeep Chakraborty, Ishita Mehta, Mugdha Kulashreshtha, and B.J. Rao

Abstract

The spatial organization of the genome within the interphase nucleus is important for mediating genome functions. The radial organization of chromosome territories has been studied traditionally using two-dimensional fluorescence in situ hybridization (FISH) using labeled whole chromosome probes. Information from 2D-FISH images is analyzed quantitatively and is depicted in the form of the spatial distribution of chromosomes territories. However, to the best of our knowledge no open-access tools are available to delineate the position of chromosome territories from 2D-FISH images. In this chapter we present a methodology termed *Im*age *A*nalysis of *C*hromosomes for comp*u*ting their *localizat*ion (IMACULAT). IMACULAT is an open-access, automated tool that partitions the cell nucleus into shells of equal area or volume and computes the spatial distribution of chromosome territories.

Key words Chromosome territories, 2D-FISH analyses tool, Nuclear positioning, Computational image processing, Structural genomics

1 Introduction

DNA has to be compacted and folded several-fold in order to confine it to the limited volume of a eukaryotic cell nucleus. Several important dynamic cellular responses are dictated by this higher order chromatin organization, and hence precision in maintaining gene accessibility and chromatin structure during the DNA folding process is essential [1]. In an interphase nucleus the genome is organized into discrete regions termed "Chromosome Territories" (CTs) [2–9]. The radial non-random organization of these CTs is conserved in vertebrates, and is greatly influenced by DNA sequence, gene density, and expression status [8, 10, 11].

The nucleus has compartmentalized its functions into subnuclear areas that are enriched in specific factors for particular processes such as transcription, replication, and repair [8, 12, 13]. Thus, placement of CTs within the nucleus influences their

Ronald Hancock (ed.), *The Nucleus*, Methods in Molecular Biology, vol. 1228,
DOI 10.1007/978-1-4939-1680-1_17, © Springer Science+Business Media New York 2015

interactions with various nuclear bodies, specific factors, and other chromosomes, thereby controlling cellular functionality [14, 15]. Interaction of a gene promoter with a regulatory region, or the presence of a chromosomal domain in proximity to a transcription factory, is known to dictate the function of those genomic elements [16–19]. Association with the nuclear periphery or the nucleolus may serve as a repressive environment for a gene locus [2, 4, 20–24]. The organization of the genome within an interphase nucleus is dynamic and an alteration in localization of gene loci, specific chromosomal domains, and whole CTs has been reported in response to external stimuli such as differentiation, disease, and stress [8, 21, 25–29]. For instance, when a cell incurs DNA damage chromosome 19 territories have been shown to relocate from the nuclear interior to the periphery [29]. Thus, understanding the spatial organization of the genome within the nucleus would help us to unravel the intricate molecular details of genome function.

The spatial organization of the genome has been traditionally studied using techniques such as fluorescence in situ hybridization (FISH) [30]. Since FISH on two-dimensional (flattened) cells is fast and yields higher sample numbers, it is the most prevalent technique for studying spatial organization of human chromosomes (*see* **Note 1**). A major obstacle while doing 2D-FISH is the lack of open access software that could analyze the images to deduce the spatial organization of chromosomes. In this chapter we present a methodology that quantitatively maps the location of chromosomes within an interphase nucleus, termed IMACULAT (IMage Analysis of Chromosomes for CompUting LocAlizaTion) [31]. IMACULAT partitions the nucleus into shells of equal areas or equal volumes and quantitates the amount of whole chromosome probe in each shell. Histograms are then plotted after normalization to the total amount of DNA in each shell, and thus chromosome localization is determined to be at the nuclear periphery, intermediate, or interior. The results obtained are in an Excel file and the user can check the credibility of the data by comparing the quantitative data with the corresponding image that shows the elliptical shells. Positioning all human chromosomes in primary human dermal fibroblasts was used as an assay to validate the functionality of IMACULAT. These analyses not only helped us to corroborate the spatial positions of individual CTs with previous studies [2, 31, 32], but also reinforced the gene density-based organization of CTs in proliferating human dermal fibroblasts [31]. Further, IMACULAT is automated, fast, not labor-intensive, and can handle large datasets, thereby yielding statistically significant analysis. In a recent study IMACULAT was used to identify a novel DNA damage-dependent spatial reorganization [29]. Upon induction of DNA damage, certain territories for chromosomes such as 12, 15, 17, and 19 undergo large-scale repositioning in both dermal and lung fibroblasts. After positioning these CTs at various DNA damage doses and at different time intervals, IMACULAT was used to show that this repositioning is dose-dependent, temporally regulated, and requires activity of DNA damage sensors. This study highlights the importance of a methodology such as IMACULAT to identify a novel relationship between nuclear architecture and a cellular response [29].

2 Materials

All solutions should be prepared in ultrapure double-distilled water (deionized water purified to attain a sensitivity of 18 MΩ-cm at 25 °C). Ensure that all waste is disposed off appropriately.

2.1 Cell Culture

Store all reagents at –20 °C before making complete media, unless indicated otherwise. After addition of solutions to media, store them at 4 °C and use within 1 month. Ensure that all media components are maintained in sterile conditions.

1. Primary normal human dermal fibroblasts (Lonza).

2. Complete growth medium: Dulbecco's modified Eagle's Medium (DMEM) (Gibco) supplemented with 15 % (v/v) fetal bovine serum (Gibco), 2 % (v/v) antibiotic-antimycotic (Gibco), and 200 mM L-glutamine (Gibco).

3. 0.25 % trypsin solution (Gibco) (*see* **Note 2**).

4. Versene solution: 0.8 % (w/v) NaCl, 0.02 % (w/v) KCl, 0.0115 % (w/v) Na_2HPO_4, 0.02 % (w/v) KH_2PO_4, 0.2 % (w/v) EDTA. Store at 4 °C.

5. Hemocytometer (Neubauer, $1/400$ mm^2) with a chamber depth of 0.1 mm.

6. Incubator at 37 °C with 5 % CO_2.

7. Phase-contrast microscope.

8. Benchtop centrifuge.

2.2 Fluorescence In Situ Hybridization (2D-FISH)

1. 0.075 M KCl solution.

2. Fresh ice-cold methanol–acetic acid (3:1).

3. Benchtop centrifuge.

4. SuperFrost slides.

5. 20× SSC: 3 M NaCl, 300 mM Na_3 citrate.$2H_2O$, pH 7.0.

6. 70 % (v/v) formamide, 2× SSC (*see* **Note 3**).

7. Water baths at 70 and 37 °C.

8. Thermostated plate.

9. Heating block.

10. 70, 80, and 100 % ethanol solutions.

11. Human whole chromosome probe (Applied Spectral Imaging, Carlsbad, CA, USA).

12. Rubber cement.

13. Tween-20.

14. Mounting medium: VECTASHIELD (Vector Laboratories) containing DAPI (2 μg/mL).

15. Microscope: Zeiss Axiovert 200 with Axiovision software, or similar.

3 Methods

3.1 Cell Culture

Carry out all the steps in a sterile environment at room temperature. Pre-warm all reagents used for cell culture to 37 °C.

1. Maintain primary normal human dermal fibroblasts in complete DMEM medium and passage them twice weekly. Wash the cells with versene and then treat them with trypsin (5 mL) to detach them from the plastic or glass surface (see **Note 2**).

2. Once the cells are detached from the surface, collect them in a 15 mL Falcon centrifuge tube. Wash the flask with 5 mL of growth medium and add that medium to the tube containing the cells. Centrifuge the cell suspension at $300–400 \times g$ for 5 min. Remove the supernatant and resuspend the cells in a known volume of fresh growth medium.

3. Estimate the number of cells in the suspension by placing an aliquot on a hemocytometer. Use the following equation to estimate the number of cells:

$$\frac{Number\,of\,cells}{Number\,of\,squares} \times Volume\,of\,cell\,suspension \times 10^4$$

Seed the cells in culture dishes at a density of 3×10^3 per cm^2 area. Maintain the cells at 37 °C in a humidified atmosphere containing 5 % CO_2.

3.2 2D-Fluorescence In Situ Hybridization (2D-FISH)

Perform a standard 2D-FISH assay [33].

1. Before fixation, culture the cells for at least 2 days after passaging.

2. Harvest the cells in trypsin (see **steps 1** and **2** above) and resuspend them in 0.075 M KCl (see **Note 4**). Leave the cells in this hypotonic solution at room temperature for 15 min.

3. Centrifuge the cell suspension at $300 \times g$ for 5 min and discard the supernatant.

4. Resuspend the cell pellet by drop-wise addition of freshly made ice-cold methanol–acetic acid (3:1) with gentle shaking. Incubate the cells in this fixative for at least 1 h up to a maximum of 18 h at 4 °C.

5. Repeat the above fixation and centrifugation steps at least 4–5 times, but only leaving the cells on ice for 5–15 min. The fixed cell suspension can be left at 4 or –20 °C until use.

6. Prewash the slides with 70 % ethanol.

7. Centrifuge the cell suspension and resuspend the pellet in a small amount of fresh ice-cold fixative. Drop about 20–50 µL of cell suspension onto slides from a height to spread out the nuclei.

8. Allow the slides to air-dry and visualize them under the microscope to ensure that the density of cells is suitable.

9. Dip the slides through an ethanol sequence (70, 80, and 100 %) for 3 min each and then air-dry.

10. Denature the slides by plunging them into 70 % formamide/2× SSC pre-warmed at 70 °C (*see* **Note 3**) for precisely 2 min, immediately place them in ice-cold 70 % ethanol for 3 min, and then take them through 90 and 100 % ethanol for 3 min each.

11. Dry the slides at 37 °C. Slides are now ready for hybridization with the probe.

12. Denature the whole chromosome human probe according to the supplier's protocol by heating at 80 °C for 7 min followed by renaturation at 37 °C for at least 30 min.

13. Place 30 µL of probe solution on the cells, place a coverslip, and seal using rubber cement. Allow the slides to hybridize with the probe at 37 °C overnight (16–24 h).

14. Remove the rubber cement from the coverslip carefully and detach the coverslip from the slide. Wash the slides in 0.4× SSC and 4× SSC containing 0.1 % Tween-20 (v/v) for 3 min each (thrice in each buffer) at 72 °C.

15. Mount the slides in VECTASHIELD mounting media containing DAPI (2 µg/mL).

16. Visualize the slides using a 100× oil immersion lens on a Zeiss Axiovert 200 or similar microscope. Capture at least 100 random images per sample using a cooled CCD camera, following a rectangular scan pattern.

17. Pseudo-color and merge the images using Axiovision software.

3.3 Image Processing

All images are run through the IMACULAT area module which divides the nuclei into five concentric shells of equal area, and the IMACULAT volume module that divides the nuclei into five concentric shells of areas such that the 3D volumes of the shells is

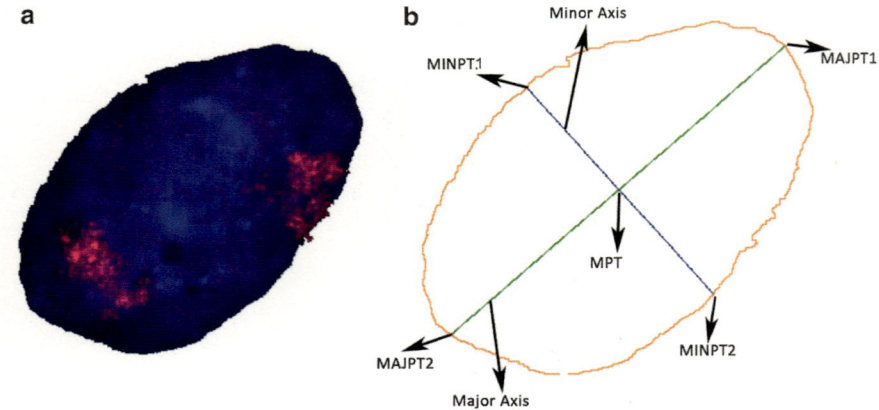

Fig. 1 Determining the major and the minor axes of nuclei using IMACULAT. (**a**) The contour of the nucleus (in *blue*) and of the 2D-FISH image of a chromosome territory (in *red*) are determined and the background is colored *white*. (**b**) The contour is then colored *yellow* while the major and minor axes are in *green* and *blue*, respectively [31]. *MPT* mid-point of the nucleus, *MINPT1 and MINPT2* points of intersection of minor axes with the perimeter, *MAJPT1 and MAJPT2* points of intersection of major axes with the perimeter

equal. The IMACULAT methodology has been detailed previously [31]. In short, the periphery is determined followed by identification of the center of the ellipse and computing the major and the minor axis (Fig. 1) (*see* **Note 5**).

Subsequently, the ellipse is divided into shells such that the area of each shell is equal. This processing, referred to as equal area portioning, does not result in an equal volume portioning of the nucleus (due to the cubic relation of the radii to the volume). In order to obtain an equal volume shell distribution, the areas of the 2D ellipse are partitioned in the ratio of 34, 20, 17, 16, and 13 (from the innermost to the outermost shell) (Fig. 2).

The difference between the equal area and equal volume analyses is highlighted by the slight shift in the slope change; equal volume analyses results in detection of less chromatin in the outermost shell and more chromatin in the innermost shell, thereby dampening the slope compared to equal area analyses (Fig. 3). However, the results obtained from both equal area and equal volume analyses display similar distribution trends and can indeed detect a distinct change in distribution (interior to periphery or vice versa).

The installation procedure for the IMACULAT program can be accessed at http://www.sanchak.com/install.html. In order to determine the spatial location of a chromosome, the images should be stored in a directory (*see* **Note 6**). The runtime of the IMACULAT program depends on the image size, with higher resolution images requiring more computational power (*see* **Note 7**).

Equal Area Analyses   Equal Volume Analyses

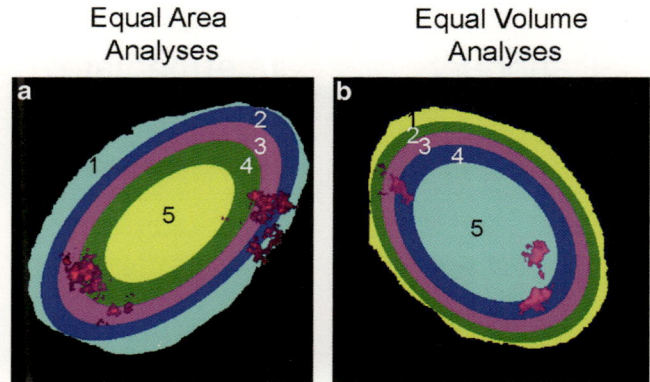

Fig. 2 Portioning the nucleus into shells. The nucleus is partitioned into five concentric shells of equal area (**a**) or equal volume (**b**) starting from the center. The percentage of the chromosome signal in each shell is then determined by calculating the number of red pixels in each shell [29]

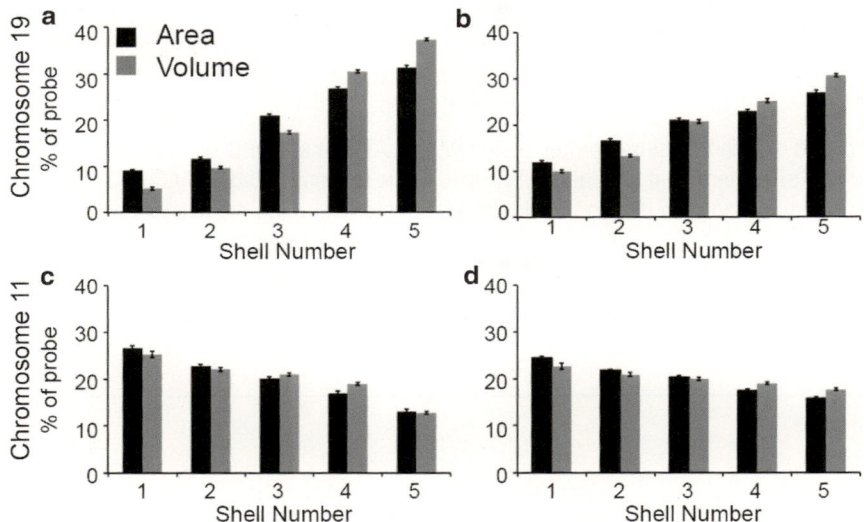

Fig. 3 The differences between equal area and equal volume analyses. Equal volume analyses results in less chromatin in the peripheral shell and more in the nuclear interior, as compared to equal area analyses. This causes a slight shift in the slope of distribution of chromosomes (Panel (**a**) and (**b**), chromosome 19; Panel (**c**) and (**d**), chromosome 11) and is evident by comparing the slopes of *black* (equal area analyses) and *grey* (equal volume analyses) bars [30]

IMACULAT measures the total signal from DAPI and from the probe in each of the five shells, and then normalizes the probe signal to the DAPI signal. The normalized amount of probe in each shell is then plotted as a histogram with error bars representing the standard error of the mean (Fig. 4). Statistical tests such as ANOVA or t-tests are performed to compute significant differences in the test vs. control sets.

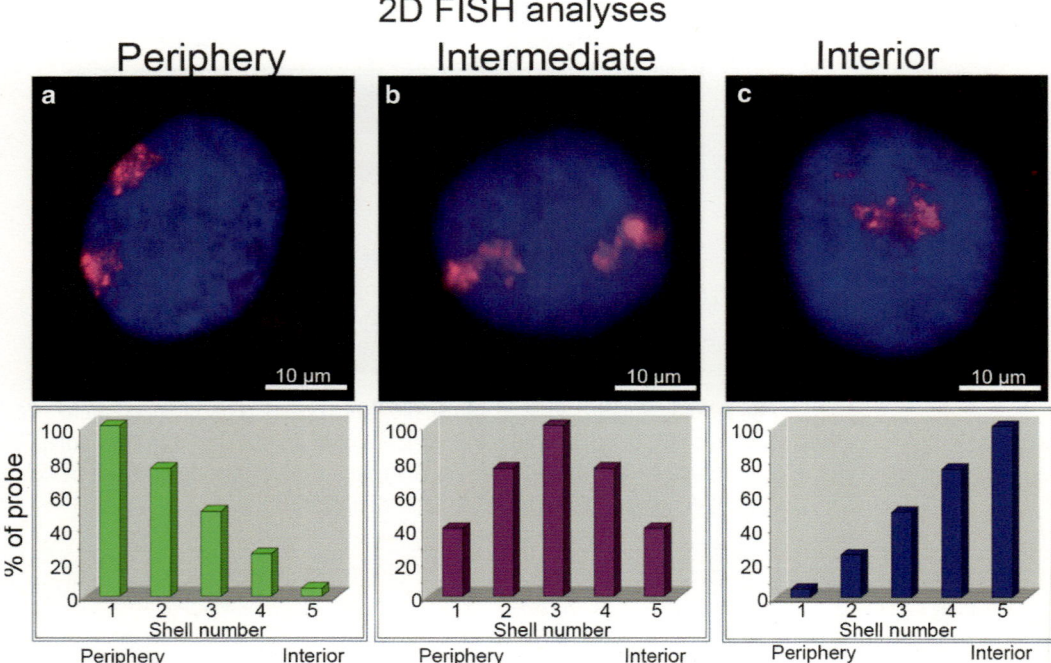

Fig. 4 Distribution of a chromosome territory using IMACULAT. The amounts of chromosome probe in the five concentric shells are determined and then normalized to the amount of DAPI. IMACULAT output is in the form of an Excel file, which can then be used to plot histograms to determine the distribution of chromosome territories. Panel (**a**) shows a typical histogram displayed by a peripherally localized CT, while a CT in the nuclear interior shows a distribution as in panel (**c**). A *bell-shaped curve* is usually a hallmark of a chromosome that occupies an intermediate location in the nucleus (panel **b**) [29]

4   Notes

1. 2D-FISH involves flattening of the cell nuclei, resulting in a bias with higher DAPI signal in the nuclear center compared to the periphery, thus making this a less accurate technique. With modern microscopy and advances in FISH, it is definitely more accurate to delineate CTs using three-dimensional FISH. However, this technique is time-consuming and labor-intensive, which makes it difficult to employ for global genome studies. We recommend that all major findings obtained using IMACULAT be confirmed using manual 3D-FISH analyses.

2. Trypsin is known to induce stress in cells if used at excessive concentrations. While passaging the cells, do not treat the cells with trypsin for more than 5 min.

3. Formamide is highly corrosive on contact with skin or eyes and may be deadly if ingested. It is also a teratogen and hence its

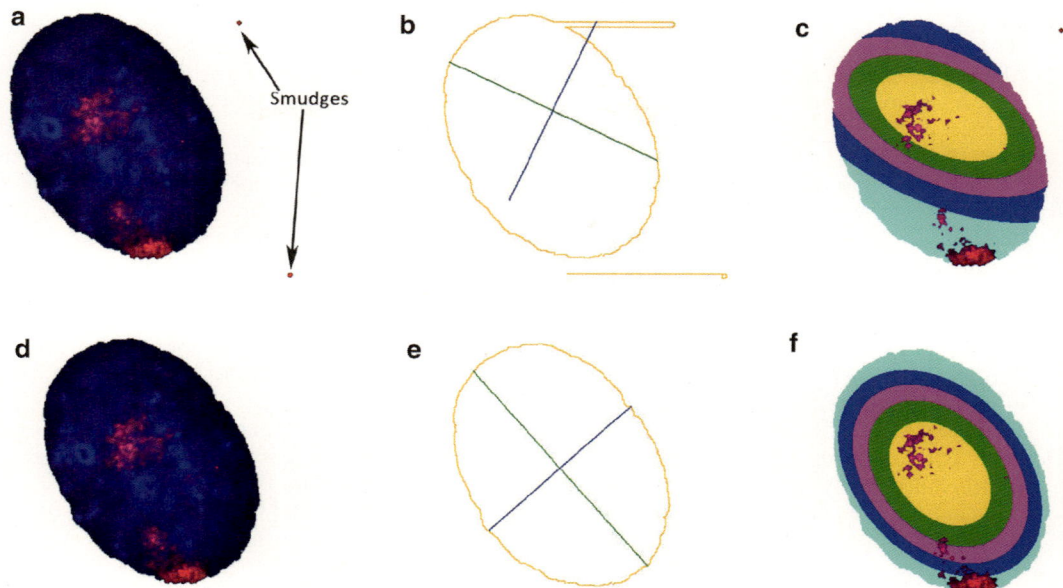

Fig. 5 Background removal for IMACULAT processing. (**a**) Original 2D-FISH images may have some background ("smudges") outside the nuclear contour (marked in *red*). (**b**) Processing a nucleus with such a background leads to erroneous determination of the major and minor axes. (**c**) Consequently, the partitioning of the nucleus is also incorrect. (**d**) The pre-processing step removes background smudges. (**e**) The computation of the contour and of the major and minor axes and the partitioning are now correct (**f**) [31]

use should be avoided in cases of pregnancy. Do not touch formamide without wearing gloves.

4. Addition of KCl and fixative to the cells should be done drop-wise whilst mixing continuously, in order to avoid clumping.

5. Before analyzing an image by IMACULAT, remove any excess background. Background may affect the ability of IMACULAT to identify the nuclear contour, leading to erroneous detection of the major and minor axes (Fig. 5).

6. When saving images to be analyzed by IMACULAT, ensure that they do not have any spaces in their names. This is primarily due to the UNIX OS restriction.

7. Images to be processed through IMACULAT should ideally have a resolution of 512×512 pixels. The runtime of IMACULAT is directly proportional to the image size and the number of images to be processed. The resolution of 512×512 pixels ensures an optimal runtime without losing any critical information about the images.

References

1. Alberts B, Lewis J, Raff M et al (2007) Molecular biology of the cell, 5th edn. Garland Science, New York

2. Boyle S, Gilchrist S, Bridger JM et al (2001) The spatial organization of human chromosomes within the nuclei of normal and emerin-mutant cells. Hum Mol Genet 10:211–219

3. Cremer C, Zorn C, Cremer T (1974) An ultraviolet laser microbeam for 257 nm. Microsc Acta 75:331–337

4. Croft JA, Bridger JM, Boyle S et al (1999) Differences in the localization and morphology of chromosomes in the human nucleus. J Cell Biol 145:1119–1131

5. Zorn C, Cremer C, Cremer T et al (1979) Unscheduled DNA synthesis after partial UV irradiation of the cell nucleus. Distribution in interphase and metaphase. Exp Cell Res 124:111–119

6. Meaburn KJ, Misteli T (2007) Cell biology: chromosome territories. Nature 445:379–781

7. Parada L, Misteli T (2002) Chromosome positioning in the interphase nucleus. Trends Cell Biol 12:425–432

8. Foster HA, Bridger JM (2005) The genome and the nucleus: a marriage made by evolution. Chromosoma 114:212–229

9. Cremer T, Cremer C (2001) Chromosome territories, nuclear architecture and gene regulation in mammalian cells. Nat Rev Genet 2:292–301

10. Tanabe H, Muller S, Neusser M et al (2002) Evolutionary conservation of chromosome territory arrangements in cell nuclei from higher primates. Proc Natl Acad Sci U S A 99:4424–4429

11. Foster HA, Griffin DK et al (2012) Interphase chromosome positioning in in vitro porcine cells and ex vivo porcine tissues. BMC Cell Biol 13:30

12. Strouboulis J, Wolffe AP (1996) Functional compartmentalization of the nucleus. J Cell Sci 109:1991–2000

13. van Driel R, Humbel B, de Jong L (1991) The nucleus: a black box being opened. J Cell Biochem 47:311–316

14. Fraser P, Bickmore W (2007) Nuclear organization of the genome and the potential for gene regulation. Nature 447:413–417

15. Bickmore WA, van Steensel B (2013) Genome architecture: domain organization of interphase chromosomes. Cell 152:1270–1284

16. Brown JM, Leach J, Reittie JE et al (2006) Coregulated human globin genes are frequently in spatial proximity when active. J Cell Biol 172:177–187

17. Eils R, Dietzel S, Bertin E et al (1996) Three-dimensional reconstruction of painted human interphase chromosomes: active and inactive X chromosome territories have similar volumes but differ in shape and surface structure. J Cell Biol 135:1427–1440

18. Chambeyron S, Bickmore WA (2004) Chromatin decondensation and nuclear reorganization of the HoxB locus upon induction of transcription. Genes Dev 18:1119–1130

19. Ferrai C, Xie SQ, Luraghi P et al (2010) Poised transcription factories prime silent uPA gene prior to activation. PLoS Biol 8:e1000270

20. Chubb JR, Boyle S, Perry P et al (2002) Chromatin motion is constrained by association with nuclear compartments in human cells. Curr Biol 12:439–445

21. Kim SH, McQueen PG, Lichtman MK et al (2004) Spatial genome organization during T-cell differentiation. Cytogenet Genome Res 105:292–301

22. Hewitt SL, High FA, Reiner SL et al (2004) Nuclear repositioning marks the selective exclusion of lineage-inappropriate transcription factor loci during T helper cell differentiation. Eur J Immunol 34:3604–3613

23. Zink D, Amaral MD, Englmann A et al (2004) Transcription-dependent spatial arrangements of CFTR and adjacent genes in human cell nuclei. J Cell Biol 166:815–825

24. Gilbert N, Gilchrist S, Bickmore WA (2005) Chromatin organization in the mammalian nucleus. Int Rev Cytol 242:283–336

25. Casolari JM, Brown CR, Drubin DA et al (2005) Developmentally induced changes in transcriptional program alter spatial organization across chromosomes. Genes Dev 19:1188–1198

26. Kuroda M, Tanabe H, Yoshida K et al (2004) Alteration of chromosome positioning during adipocyte differentiation. J Cell Sci 117:5897–5903

27. Panning MM, Gilbert DM (2005) Spatio-temporal organization of DNA replication in murine embryonic stem, primary, and immortalized cells. J Cell Biochem 95:74–82

28. Parada LA, McQueen PG, Misteli T (2004) Tissue-specific spatial organization of genomes. Genome Biol 5:R44

29. Mehta IS, Kulashreshtha M, Chakraborty S et al (2013) Chromosome territories reposition

during DNA damage-repair response. Genome Biol 14:R135

30. Volpi EV, Bridger JM (2008) FISH glossary: an overview of the fluorescence in situ hybridization technique. Biotechniques 45:385–386

31. Mehta I, Chakraborty S, Rao BJ (2013) IMACULAT – an open access package for the quantitative analysis of chromosome localization in the nucleus. PLoS One 8:e61386

32. Mehta IS, Amira M, Harvey AJ et al (2010) Rapid chromosome territory relocation by nuclear motor activity in response to serum removal in primary human fibroblasts. Genome Biol 11:R5

33. Bridger JM, Herrmann H, Munkel C et al (1998) Identification of an interchromosomal compartment by polymerization of nuclear-targeted vimentin. J Cell Sci 111:1241–1253

INDEX

Ronald Hancock (ed.), *The Nucleus*, Methods in Molecular Biology, vol. 1228,
DOI 10.1007/978-1-4939-1680-1, © Springer Science+Business Media New York 2015

Printed by Printforce, the Netherlands